Uranus

Uranus

Esoteric & Mundane

John Townley

Samuel Weiser   New York

First published 1978

Samuel Weiser, Inc.
740 Broadway
New York, N.Y. 10003

Copyright © 1978
John Townley

All rights reserved. No part of this publication may be reproduced, stored in a retrieval system, or transmitted in any form or by any means, electronic, mechanical, photocopying, recording or otherwise, without prior permission of the copyright owner.

ISBN 0-87728-415-6

Printed in the USA
Noble Offset Printers, Inc.
New York City

*To Neal John,
     because it's his planet*

CONTENTS

I	Uranus, the Planet	13
II	Uranus, the Myth	20
III	Uranus and Religion	30
IV	Uranus, Science and Technology	37
V	Uranus, Self-understanding and Psychology	43
VI	Uranus and Creativity	49
VII	Uranus, Travel and Adventure	54
VIII	Uranus and Humor	59
IX	Uranus and the Bizarre	64
X	Uranus and Sexual Perversion	70
XI	Uranus and Crime	77
XII	Uranus and Revolution	82
XIII	Uranus and War	88
XIV	Uranus and Astrology	93
XV	Uranus in the Horoscope	98
XVI	The Discovery Chart	133
XVII	Uranus in Esoteric Astrology	150

LIST OF CHARTS

Helena Blavatsky	34
Thomas A. Edison	40
Sigmund Freud	46
Walt Disney	50
Rudyard Kipling	56
George Bernard Shaw	62
Robert Harrison	68
Terry Kolb	72
Arthur Conan Doyle	80
Thomas Jefferson	84
Napoleon	90
C.E.O. Carter	94
Discovery of Uranus	132

I
Uranus — The Planet

The second Tuesday of March in 1781 found the skies over England clear and ideal for a late evening stroller or merry pub crawler on his way home from a few pints at the local taverns.

In America there was a war raging, in a fitful sort of way, as colonial insurgents tried to throw off what they considered to be the tyrannical yoke of King George III. And, thanks to increased aid from the army and navy of England's rival France, the tide of what had been a losing war for the colonists was beginning to turn in their favor.

In France the seeds of discontent were germinating and the age old monarchy so regally brought to flower by the "Sun King" Louis IV was in its last days. In only eight years Bastille Day would catapult France headlong into more than a generation of bloody revolution and world war.

But in England all was quiet that night as a professional musician whose hobby was making telescopes and observing the heavens climbed the stairs to his homemade observatory.

His name was Frederick Wilhelm Herschel and as he waded into his forties he could be thankful that moving to England from his native Hanover had increased his fortunes and given him a comfortable existence as an arranger, composer, and choirmaster in London's flourishing music business. He was many years from his youth when he had actually marched his way through withering enemy fire with not a gun but a musical instrument in his hand. After four years of serving as a military bandsman he had obtained a discharge and, at his father's advice, emigrated to England to seek more peaceful employment as a musician. There the vagaries of that fickle business had been kind to him and he settled into a regular sort of life of performing, arranging, and teaching music, leaving him plenty of time to grind lenses for telescopes and peruse the heavens, his favorite pastime.

Certainly he had little idea as he focused his 6.2", 7-ft. reflector telescope on that chilly evening of March 13th that he was about to make history. He was about to bring into the consciousness of man the planetary symbol of all the devastating revolutions and wars that were only just in the making at the time; the strife and tumult that were to characterize the next several hundred years, at least.

To his knowledge, he was only continuing a rather ambitious project for an amateur star-gazer: reviewing the catalogue of all stars of magnitude 8 and brighter. He had just recently finished a review of magnitude 4 and brighter stars and was now going back to inspect the dimmer bodies with a 227-power instrument designed for viewing binary star-systems.

But before midnight that evening he had focused in upon an uncatalogued body late in the constellation Taurus. It was quite dim and a bit fuzzy and appeared at first glance to be a comet just beginning to glow as it approached the sun from the outer reaches of the solar system. He duly noted it down and again observed it on the next clear evening three days later.

Being a careful observer and wanting to have his observations checked and confirmed by more weighty members of his field, he wrote to other astronomers and noted scientists of the day asking them to confirm his sighting of a new "comet".

Within a few months, the so-called "comet" had been recognized as something else entirely. It was not a periodic wandering mass of particles mixed with gases or even a star cluster or galaxy. It was another planet within our own solar system, the first new planet to be discovered within recorded history.

What a startling honor to fall upon an amateur astronomer! And what a problem for the scientists of the day as well. They were faced with a number of problems, one of which was what to name the new fellow-traveller.

The first, and ultimately the final, name was suggested by Herschel's friend and correspondent J.E. Bode (of Bode's Law fame). He suggested the name Uranus, because in classical Greek mythology Uranus was the father of Saturn and therefore the logical name for the planet next farther up the line. Jean Bernouilli of the Berlin Academy suggested the name *Hypercronius* based upon the same logic, as Cronus was the Greek word for the Romanized Saturn.

But Herschel was moved by more patriotic feelings. He felt that astronomers of the future should identify the planet by the period in which it was discovered and the country it was spotted from. Thus he proffered the name *Novum Sidus Georginium*, or "George's new star" after the English monarch George III. Needless to say, this struck Britain's fancy right off, and the official star catalogue soon had the object listed as "Georgian".

France would have none of this obvious English politicizing and adopted the general title of "Herschel's Planet" for want of further agreement upon a name.

Other names suggested by leading astronomers included Cybele (the wife of Saturn in mythology), Astraea (later to be attached to an asteroid), Minerva (another later asteroid), and even Neptune! Somehow, Bode's original suggestion managed to stick in the Anglo/American part of the world, although in many parts of Europe the planet is still called Herschel. Leverrier, the later discoverer of Neptune, was one enthusiastic supporter of naming the planet after its discoverer, but he may have had ulterior motives.

At any rate, by 1856 the name "Georgian" had been abandoned in British records in favor of Uranus, and the tradition of naming newly-discovered planets after gods of the Greco-Roman pantheon had been established.

There was also the question of deciding what symbol should be used for the planet in astronomical (and subsequently astrological) terminology. Bode suggested the chemical symbol for platinum, the Mars symbol standing straight up with a central dot added (♁). That is still the most common symbol for the planet outside of the United States.

But it was French astronomer Joseph Lalande's seemingly compromise symbol incorporating Herschel's own initial that has become accepted here (⛢). It is perhaps well that the discoverer should live on in the symbol, even though few know his name today.

It took many years after Herschel's first sighting for astronomers to get any kind of firm grip on the characteristics of the new planet. It was so dim and so far away that it defied conclusive analysis concerning its makeup. It still does, and will until the fly-bys of future space probes give us more concrete information on the planet's constitution.

In fact, the fifth satellite of Uranus was only discovered as late as 1948. Perhaps there are more, as yet invisible to terrestial telescopes.

But what is clear about Uranus, from an astronomical point of view, is that it is unusual. Other planets, similar to the earth, have their north/south polar axis more or less perpendicular to their orbits. Uranus, however, has its poles less than 8 degrees off its orbital plane. In fact, its north pole points slightly south (in earth terms) of its orbit. Thus its "summer" (the temperature averages -185° Centigrade) lasts for about half its 84 year revolutionary period—and technically it defies any earth-oriented definition of seasons at all. In earth terms, Uranus seems to be half standing on its head, the oddball planet in the solar system.

Next to earth and the other inner high-density planets, Uranus is fairly light-weight. Its density is only 1.71 that of water. But, like the other giant outer planets, its size makes up for it, being 14.7 times the mass of earth, and 47 times its volume.

Surprisingly, while it takes our planet 24 hours to spin once around, Uranus makes its rotation in only 10 hours 49 minutes, a dizzying rate for such a large planet.

Its diameter is 29,300 miles but what's inside that mighty girth is somewhat of a mystery. Uranus, like Saturn, Jupiter, and Neptune appears from spectral analysis to be made of mostly ammonia and methane. Its actual makeup may be yet to be discovered, as recent observations of Jupiter by space probes and airborne telescopes have revealed it to be rather different than telescopic observations had suggested. One such observation is that Uranus, like Saturn, appears to have a band of rings encircling it, first observed during a stellar occultation. Their origin, like those of Saturn, is as yet unknown, and there is no reason to suggest that further investigations of Uranus will not reveal similar anomalies.

Some currently available statistics on Uranus:

Inclination of North Pole to ecliptic—7°53′ South
Magnitude—varies from 5.7 to 6.3
Orbital velocity—4.22 miles per second
Distance from Sun—1,782,000,000 miles
Rotational period—10 hrs., 49 min.
Density—1.17 that of water
Diameter—29,300 miles
Mass—14.7 that of the earth
Volume—47 times that of earth
Gravity—1.12 that of earth
Temperature (surface)— -185°C and less
Length of year—84.01331 tropical earth years

But what is not known about Uranus are its most important characteristics: its origin, how it got to be where it is, how its polar orientation became so out of line with the rest of the planets. Here, again, future space probes will give us a better account of its history. Until then, Uranus will continue to be the mysterious renegade of the solar system.

Herschel, of course, suspected none of this. He only knew that he had sighted a new body in the skies, one of many comets, asteroids, and other chunks of space matter that were turning up at the time. It turned out to be a lot more than he reckoned, however. He became one of only three sky-gazers so far to locate a new planet not visible to the naked eye. Quite an achievement for a modest choirmaster, retired from the military, whose hobby was astronomy.

For astrology, something far from Herschel's mind and experience, it was a spectacular discovery as well. It marked a new body to be placed in all previous horoscopes and examined and analyzed. Eventually, it became evident that it was a planet that typified and represented much of the values that the age in which it was discovered expressed. It was to be acknowledged as the planet of revolution; of clear and unbiased scientific discovery. Its discovery marked the beginning of social, political, and technological systems that were to change irrevocably the structure of man's directions and efforts. And recognition of its existence brought into man's consciousness the beginnings of that so-called "Age of Aquarius" (which Uranus rules) that we are now entering as a result of the social and technological changes begun at the time of Uranus' discovery.

When a middle-aged musician and amateur astronomer stepped up to his telescope on the evening of March 13, 1781, he was about to change the world. The total efforts of most of the rest of his life—his musical compositions, arrangements, performances, and all the other primary efforts he attempted in order to make a mark upon society—have been lost or forgotten.

And had he chosen to go and have a pint with the boys at the local pub that night, he would have faded into the obscurity of history as do most men. A few mugs of good English ale could have lulled him to sleep along with the majority of loyal Englishmen.

But his fascination with the heavens, combined with fortunate atmospheric conditions, led him to his roof that night and forever tied his name into the history of a science he enjoyed but never thought to find importance in. He became the accidental discoverer of a planet. This achievement was never before accomplished and it was a planet to become associated with the very kind of unexpected accident by which it was discovered.

A new planet had been discovered: the so-called Age of Discovery had been opened. Astronomers would make every effort through observation and calculation to find more unknown planets circling the sun, but none would ever simply step onto his roof and have the solar system walk in on him like Herschel did. The very symbolism of Uranus had made itself forever unique in the circumstance of its discovery.

II
Uranus—The Myth

All the names of the planets derive from the pantheon of ancient Greece and Rome with the exception of Earth itself, whose origin is Teutonic.

The names of the five planets which are visible to the naked eye go back into prehistory. The natures of the planets in astrological lore and the gods they were associated with grew and developed hand in hand. Thus, an inspection of the myths concerning those gods provides invaluable insight into their astrological natures.

The three outer planets are also named after gods from the same pantheon but they were not named with any intentional reference to possible astrological significance. They were named by astronomers for reasons of their own, which reasons would hardly be astrological.

It would therefore seem safe to assume that there would be little likely connection between the astrological meanings of these planets and the myths surrounding the names they received.

In large measure, however, that is not the case. There are, in fact, many close parallels between the myths and their corresponding planets. Much insight into their meanings may be gleaned from a careful analysis of their myths.

This might come somewhat from early astrologers making such attributions arbitrarily once the planets were named, but that does not seem likely. The use of the name Uranus did not become widely popular, much less officially accepted, until well after astrologers had established its basic significance in the birth chart using the name Herschel, which many Europeans still call it today.

Neptune's name gained quick acceptance. Pluto was mentioned in one astrological textbook by name as early as 1910, and referred to as an unknown outer planet. This was twenty years before it was discovered! Astronomers finally chose that name completely unaware that it was already in circulation for more than a generation among astrologers.

Thus we may surmise that things have a certain way of naturally falling into place, at least in these affairs. Whatever the nature or motives of the namer, and in Pluto's case it was a postcard from a 9-year-old English girl that suggested the name to its discoverers, the right name will come to pass—witness the fortunate demise of the title "Georgian" for Herschel's discovery.

As we inspect the myth of the Greek god Uranus, this is what we find:

In the ancient Greek cosmology, the world began as Chaos, which is self-explanatory. It did not start with a void out of which a divine will forged things as in the Judaeo-Christian version, but just a jumble of disconnected existence with no order whatsoever.

From out of Chaos came Gaea, the earth, the primal protean Mother of everything. By what process this occurred is not known or at least not mentioned by classic authors such as Hesiod, whose *Theogony* gives the most extensive accounts of the early Greek pantheon. But it is not really necessary to invent a specific method of bringing this female order out of Chaos, we may just rest content to know that mothers have a way of doing that sort of thing.

Gaea then singlehandedly gave birth to Uranus, as there was no one else around to help her with the procreation. Uranus became the heavens, or "Ouranois" in Greek. He was the Sky-god and his name became a part of the Greek language in a mundane as well as mythical sense. "Ouranois" simply meant the heavens and is, in fact, the word used in the original Greek of the Lord's Prayer: "Our Father, Who art in the heavens" ("Ouranois" is the plural form, of which the singular is never used when referring to the Great Upstairs).

Uranus wrapped himself around his mother Gaea as the sky indeed must, and from this seemingly incestuous union (there was no one else around) there sprang the Titans: Oceanus, Iapetus, Cronus, the three one-eyed Cyclops, and the three 100-handed Giants.

Uranus was filled with fear and loathing at his monstrous progeny and promptly locked them away in the great abyss of Tartarus. Tartarus was the lowest place in the universe, being as far below Hades as earth was from heaven. An object falling from earth would take nine days to get there; in modern terms, that's a good many light-years. To make sure that whoever got put there, stayed, there was in Tartarus a triple wall of bronze surrounded by the burning stream Phlegethon. As far as the Titans were concerned, this made Alcatraz look like a playground.

This mistreatment of her children angered Gaea and she fomented revolt among the Titans. Only Cronus, (the Roman Saturn) the youngest, had the courage to rebel against his father. So Gaea hid him, and when Uranus came to lay with Gaea in the evening, Cronus leaped at him brandishing a sickle and castrated him with it, flinging the severed member into the sea.

He then proceeded to dismember his father and fling the pieces about with abandon.

Gaea, however, collected the gory remains, and from Uranus' blood sprang forth the Furies, the Giants, and the Tree Nymphs. The member Cronus flung into the sea united with the foam and from it arose Aphrodite, the goddess of love and beauty.

Uranus did not go unavenged in the long run, however. Cronus married his sister Rhea and, remembering what he had done to his own father, promptly ate all his own progeny at birth. Only Zeus (the Roman Jupiter) escaped when Rhea substituted a stone wrapped in swaddling clothes for Cronus' latest neonate meal. She secreted him away with the help of nymphs to Crete where, when he reached maturity, he confronted his father and forced him to disgorge his brothers and sisters who were apparently still in pretty good shape, as they went on to comprise six of the gods on Olympus, no small feat.

Thence ensued a battle between Zeus, leading the Olympic gods, and the remaining Titans led by Cronus, with only Oceanus not participating on the Titan side. He had more peaceful motives, being content to flow gently around the skirts of Gaea, providing the boundaries between heaven, earth, and Hades.

Zeus was determined to be rid of the monstrous Titans for good, but it was only through employing the help of the three 100-handed giants that he succeeded in carrying the day. He then cast the Titans down into Tartarus again, stationing the 100-handed giants as guards where they remain to this day. Some versions of the legend have it that Cronus alone was permitted exile, a la Bonaparte, in the Blessed Isles of the West.

Zeus then acceded to the throne of the Sky-god, having avenged his grandfather Uranus and thus established the universal order of the Olympian gods under which classical civilization was later to flourish.

This myth is not unique to the Greeks at all, and may in fact have been simply adopted from the earlier Hurrian culture that dominated Western Asia in the 14th and 15th centuries B.C. Their myth of Anu, the Sky-god, being castrated by his son is nearly identical to that of Uranus in many ways and certainly historically precedes it.

The basic imagery is native to myths from many other cultures as divergent as that of India and Celts, both of which had rituals of the slaying of the Father-King when he lost his virility and thus endangered the fertility of mother earth and her vitally abundant crops.

The myth has much subconscious sexual significance, as Freud was the first to point out. The castration myth represents the boy-child proving his virility and stepping into manhood by overcoming his father, though not to marry his mother as in the Oedipus myth.

Further cultural meaning can be found in the monsters springing forth from Uranus' incestuous relation with his mother and his eventual destruction as a result. The transgression of the universal incest taboo not only predictably brings forth monstrous offspring, but results in actual castration and dismemberment as well.

Even more important as a social myth is the final overthrow of the primordial Titans by the more stable and civilized gods of Olympus. Through this great struggle with the untamed savage forces of the primitive universe, the obviously man-identified Olympians finally establish order and reason. The disturbing and powerful forces of the Id, in Freudian terms, have been brought under control by the Super-ego. One could stretch the point further by identifying the primary oral stage of human development with Gaea/Uranus, the anal retentive stage with Cronus, and the final balanced genital stage with Zeus.

Certainly the Uranus myth powerfully symbolizes the bringing of order from chaos and protean struggle, but how is this reflected in the astrological attributes of the planet Uranus?

First, Uranus is the primary mythological male figure. It is proto-male with no female characteristics whatsoever. This fits with the dry, unbending qualities that are attributed astrologically to the planet. Uranus represents the brilliant, clear, harsh Truth, softened by nothing gentle, gradual, or in anyway *yin* or female. It is the ultimate *yang* principle, rather than the more human and ego-aggressive maleness that Mars represents. It is, in fact, not aggressive at all but simply, overwhelmingly stark and irresistable by its pure unmitigation. It brooks no compromise or meeting of minds, none is acceptable. The Total Truth requires no altering or embellishment. The purest nature of Uranus is rather like the more obscure concept of Abraxas: total awareness of ultimate Truth so blinding and all-encompassing that to taste of it would be sure destruction for a mere mortal. Only the highest initiate masters could bear to gaze upon that awesome aspect of God.

Similarly, Uranus is associated with brilliant sparks of inspiration or religious illumination. It is the instantaneous knowledge of God through direct revelation rather than through faith.

On a more mundane level, this quality of sudden awareness explains its connection with inventions in general, which come to the inventor in a flash. Or, on a higher level, it represents science because that field is the attempt to discover not what seems to be in the world, but the clear, irrevocable, and tangible truth of the way things provably *are*.

The association of Uranus with perversion and the sexually bizarre is obvious. His own incest led to monstrous and deformed offspring as well, another area associated with the planet Uranus (sideshow freaks and human oddities in general).

So also Uranus is connected with surprise and its offshoot, humor. A combination of the unusual with the unexpected is bound to arise when the qualities of startling truth mixed with the bizarre are revealed.

In fact, the very essence of the bizarre requires that it must be the actual truth—otherwise it would be mere fantasy and thus lose all impact. These are strictly Uranus qualities, both of myth and planet.

Revolutions and sudden radical changes are also associated with Uranus, and this is quite predictable. First, it was Uranus who first met with revolution and, indeed, brought it upon himself.

More importantly, though, when Uranus in its pure form of total truth meets resistance, it cannot compromise or blend. It must totally destroy its resistance or be destroyed itself. Certainly revolutions do not take place where compromise is available. A revolution is always the result of the conflict of two systems or realities so incompatible that one must inevitably be destroyed. To each side, its own system represents the truth, the end-all, and thus must triumph unaltered or perish in the attempt.

In similar manner, Uranus when harshly aspected is associated with crime and violence. The so-called "criminal" of the Uranus type simply cannot compromise with society to the extent of obeying its laws when he sees the truth differently. All too often this results in crimes of violence, particularly crimes of passion or crimes of desperation. Unswerving reality has impelled the individual along an inevitable path of collision with society.

Yet how many "criminals" of earlier times were later applauded as martyrs or revolutionaries when their brand of the truth became the popular belief? Everyone has a slightly different version of the "truth" and the need to follow it. When "truths" collide we have all the war, destruction, and hate that afflictions of Uranus can bring. Indeed, if there is any central keyword to attach to Uranus it would be truth, just as illusion might be ascribed to Neptune.

Thus, many different forms of truth-seeking, particularly in physical and tangible ways, are associated with Uranus. For Uranus is associated primarily with the hard, dry, masculine truths, not the more subjective feminine ones (after all, an illusion is a truth in and of itself).

So astrology in its more scientific and less mystical forms is associated with Uranus, as is exploring, travel, and adventuring, where the goal is to uncover uncharted territories or knowledge.

But Uranus' masculinity is strictly on a non-human level and thus all too often quite scary. As a proto-god he predates the civilizing influences of man and thus rules the dry and hard to govern forces of nature like electricity and the wind.

Even on a more personal level, Uranus tends to the physical and impersonal, particularly when mixed up with the more human Mars. For instance, Uranus rules the gun, but Mars the shooting of it; Uranus rules the revolution, but Mars rules the soldiers fighting it.

As the primary, pre-human, masculine deity, Uranus rules all that has masculine properties but is not in itself, in fact, male.

Uranus is the first male to break all the rules regarding sex and family. He is associated with original sin and the devastation and destruction it begets. Uranus is also related to the resulting fertility and creativity when he is punished and brought under control: both the Furies who pursue transgressors and Aphrodite, who is the goddess of love, sprang from the blood of the slain Uranus. Once properly dealt with, the untamed powers of the wild primal god can be put to all kinds of creative uses for mankind.

As has been seen, much can be learned about the planet Uranus by going back to its social myth. In this way more insight can be gained into the pure qualities of the planet than by other lines of study. In the myth we do not have the constant admixture with other elements that we necessarily find when dealing with Uranus in the natal chart or when dealing with its transits. However, the actual hard truth of such purist speculation can only be gleaned by the careful observation of a great deal of data concerning Uranus' apparent effects in everyday life. It would be doing a disservice to the nature of the planet itself not to subject it to the most careful scientific and, indeed, Uranian scrutiny to determine where the truth of the matter lies.

Fortunately, astrology is just entering such a period after a long time of being heavily influenced by Neptune. Until very recently, astrological assertions have been based strictly upon the subjective, personal experience of individual astrologers rather than upon hard facts and testable data. This combined with the esoteric mystification of the subject in the late 19th and early 20th century have left the interpretational study of the heavens (Uranus himself) in considerable disfavor in an increasingly Uranus-oriented culture.

This is not to say there is not validity in much of the esoteric approach to astrology. Perhaps much of it represents higher truths existing on planes as yet incomprehensible to the mundane human mind, a kind of aspect of Uranus too bright to easily behold. Certainly this is what esoteric astrology purports to be and to the properly-developed mind and spirit this is just what it reveals.

But the loftiest buildings must have the most firm foundations, and astrology is only now beginning to get its factual and observational foundations constructed. Without a firm grounding in physical astrology, mankind as a whole can never make esoteric astrology more than flights of fancy for the average individual, and a lonely journey for the initiate. We must have our feet firmly on the ground before we can stare the heavens squarely in the face. Otherwise the apparition of the flame of Pure Truth, Uranus' highest level, will be less likely to bring us to enlightenment and more likely to burn us to a crisp.

III
Uranus and Religion

When speaking of religion in astrological terms, reference is usually made to the planet Neptune. It is customarily associated with the spirit, the intangible, the faith and qualities of belief that are normally associated with religion.

But even the most brief perusal of the nativities of the major religious leaders of recent history will reveal that Uranus, not Neptune, is generally the more significant planet.

This is not necessarily in contradiction to traditional astrology, for it is the body of spiritual followers who most especially typify the Neptunian qualities associated with religion. It is not often that faith, trust, and unquestioning belief are the most notable qualities of spiritual innovators. Religious leaders take up their missions usually because of a clear and uncompromising vision of the Divine which must be transmitted as lucidly as possible to the rest of the world.

It is just this flashing insight into the qualities of pure spiritual Truth that reflects Uranus at its highest manifestation. It is the clear and undeniable reality that characterizes the waking vision. The follower and the saint believe and trust because they can *feel* the truth of the faith they espouse; the religious visionary has no dependence on faith or trust because he has *seen* it in its crystalized form, in stark reality.

It is a characteristic of the recent Piscean Age that Neptune has been so wound up in religion (as well as astrology). The tenets of Christianity are based on faith and belief, unbacked by doubting Thomas's reassurance of actual touching the presence of Christ. This Christian heritage is in stark contrast to, the visionary teachings of the Old Testament where God made his presence known in the most concrete manner.

But even the Christian visionaries, while preaching faith and hope, had gotten their personal word from the Source itself. St. Paul, like so many of the older prophets, received the word in no uncertain terms: a crashing vision in the middle of the day changed him from a persecuter of Christians to a follower of Christ. He may have spent the rest of his days asking converts to *have* faith, but he certainly did not speak *from* it. He had seen the Real Thing. Certainly it is easier to believe in the power of God when you have seen it operating firsthand.

Similarly, the non-secular leaders of the Church often were those who were talking from firsthand contact with the Divine. Visions are too often associated with Neptune in astrology, and incorrectly so. A dream is certainly Neptunian, but a midday revelation with its harsh brilliance is strictly Uranian. Recent sources have tried to explain away the visions of St. Joan of Arc as mere manifestations of schizophrenia. But it is hardly likely that a schizophrenic could have stood up under the rigors of battle campaigns without faltering, even unto death by fire at the hands of the English after a long and tortuous trial; and all without any sign of madness except for a very consistent set of instructional voices.

Throughout the world, it is those individuals who are inspired by what they perceive to be clear and undeniable spiritual truth that become religious leaders. "Inspired" is even the wrong word to describe the phenomenon, "certain" is more appropriate. It takes faith to be a follower, but it takes sure knowledge (or total madness) to achieve the surety to inspire others to follow.

This is not to say that every visionary has the end-all version of what the infinite is about. It is only to suggest that such persons see a part of the whole Truth so clearly that it appears absolute and therefore utterly concrete. Thus, we find religions warring with each other, each with their bit of the Absolute and each convinced that it has the whole story.

That this Uranus quality is exceedingly important is borne out by the charts of religious leaders of the past few centuries. In the Eastern-oriented faiths we find:

Ramakrishna—Uranus ruling Ascendant, conjunct Sun

Krishnamurti—Uranus ruling Ascendant, conjunct Sun

Kirpal Singh—Uranus ruling Sun, Mercury, Moon, square Sun

Meher Baba—Uranus ruling ascendant and Venus, conjunct Moon, opposition Jupiter, square Venus

Abdul Baha—Uranus ruling Ascendant, Saturn, Neptune, conjunct Jupiter

Jinarojadasa—Uranus conjunct Moon, opposition Saturn, rules Saturn and Eighth

Interestingly enough, the countless gurus that have flooded the West in recent years show much less of this clear Uranus influence. They may or may not be sincere in their teachings, but they are not speaking from the clear visions of the universe that their earlier forebears were, even within the last century or so.

Among the leaders of recent Western sects:
Blavatsky—Uranus ruling Eighth and Ninth although Neptune rules MC, conjunct Jupiter, opposition Sun
Annie Besant—Uranus ruling Neptune and Twelfth, opposition Mercury, square Jupiter and Moon
Mary Baker Eddy—Uranus ruling Moon and Second House (emphasis on physical), conjunct Neptune and Ascendant
Swedenborg—Uranus ruling Sun, conjunct Moon, square Sun

The clarity of Uranus is evident in the teachings of such leaders, even though in some cases the inspiration may not have come directly from visions. The surety of truth and the confidence that goes with it are necessary to the successful spiritual leader if his works are to be long sustained.

In the last few years another kind of Uranus-oriented semi-religious leader has emerged with, as yet, doubtful success. It is the drug-inspired visionary whose experiences in the psychedelic world have led to apparently clear images of spiritual reality.

Certainly the results aren't yet evident of this kind of vision. For many centuries, drugs of one sort or another have been a very important source of visionary activity in many cultures, but few important religious leaders have emerged solely from such sources.

On the other hand, the leaders of the modern psychedelic movement have been under such deliberate repression and persecution because of their drug-orientation that no clear view is yet available of what they may have produced. A prophet in prison, real or imagined, can have little efficacy in today's world.

Timothy Leary (Uranus ruling Moon, conjunct Moon, square Venus) is one such example. His treatment by a disapproving society has been so

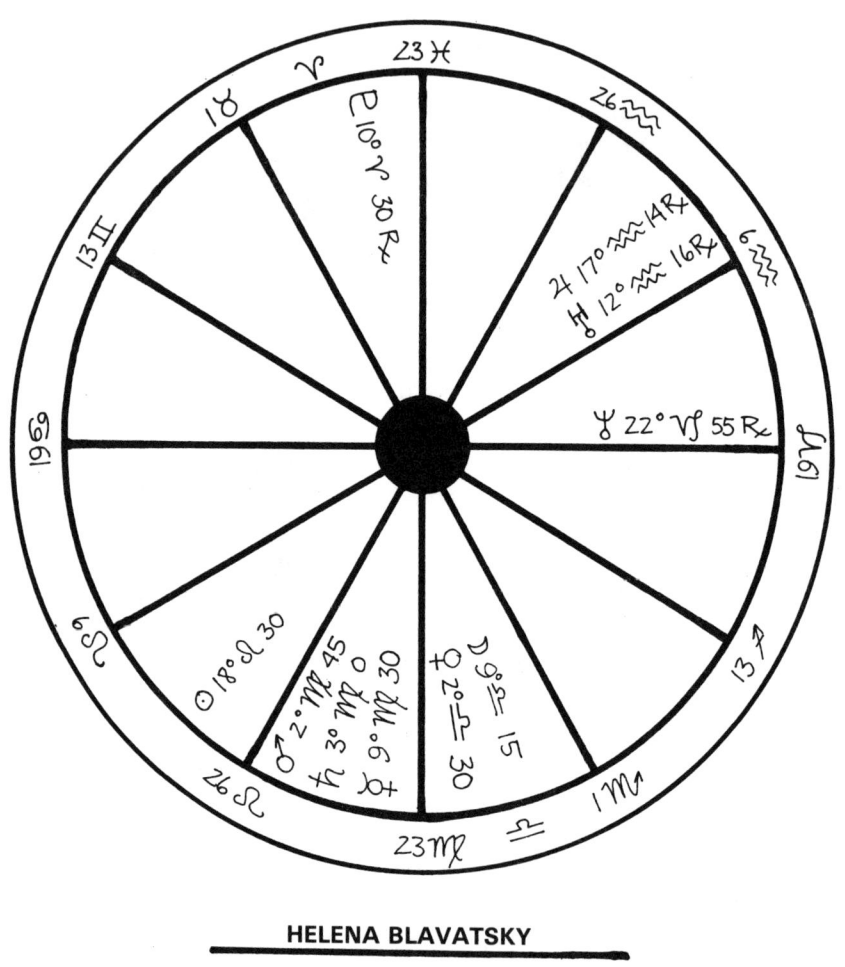

HELENA BLAVATSKY

AUGUST 12, 1831

2:17 AM

EKATERINOSLAW

extreme and repressive that it is very difficult to judge what truth his message has or how much of it we have even been allowed to hear.

Similarly, Richard Alpert (now calling himself Baba Ram Dass), Leary's psychedelic co-explorer has adopted more socially acceptable pastures for spiritual teachings following a more trendy neo-eastern path, but he still has his foundings in the visionary drug world and the results of it are still not clear—despite his Uranus conjunct Sun, square Jupiter and Pluto.

Perhaps the Uranus aspects of the 1960's drug culture will never really be made clear, thanks to the very Neptunian nature of drugs in general. Nevertheless, it is very clear that many, throughout history, have had the same kind of flashing Uranus clarity of vision with the help of drugs. Dependency upon such chemicals to produce visions, however, has never produced any sort of world religious leader. Drugs may, in many cases, suddenly open a window onto the Truth, but only those grounded in reality have so far managed to keep that window open for any length of time.

Whatever the source of the clear religious vision, it is likely to be characterized by Uranus. Uranus is that seemingly totally clear Reality that we deify in our search for the truth as Religion. In its lower forms, it provides warring forces with such clear inspiration that they destroy each other in its name. Truth has its levels, and its varieties. Many a person has died, probably unnecessarily, defending his purely-perceived piece of the greater Truth.

Throughout, on its spiritual level, Uranus represents that most real, tangible existence, the total expression of which would be completely blinding to the beholder. That is because its basic nature is separative, not blending. To see the Ultimate as a separate individual is to become overwhelmed and destroyed by it. To become and be absorbed into that same Cosmos (in the Neptunian sense) is to be realized and deified.

Thus, religious achievement (if it can be called that) in the long run depends upon the combined opposites of Uranus and Neptune. Uranus can only perceive God, the separate, the masculine, the yang. But only Neptune can become God, the unity, the feminine, the yin. Naturally, in a male-oriented society, the leaders of religion have been the Uranus-type, while the followers have been of the Neptunian type.

Perhaps in the newly-arriving Aquarian Age such roles will reverse continuing the recent grip that science has had upon the masses and Neptunian religion has had upon the more educated intellectual elite.

But whatever the social circumstances, Uranus will still retain its religious and spiritual place as the perfect beholder, the clear and visible light of specific Truth in all of its many manifestations: obsessive and destructive to the unenlightened, all-consuming and transforming to the true visionary. Whether a person is one or the other is still the puzzle. No matter how convincing, is what we clearly seem to see really the Truth?

IV
Uranus, Science and Technology

It is easier to conceive of Uranus as being connected with science and technology than with religion. In this current age, we tend to think of science rather than religion as the precise, physical truth of matters. Everyone can see undeniable evidence for the law of gravity, while it is hard to get any kind of grip upon the varying versions of theology, unless one has actually experienced a waking vision.

Seeing is believing, and science, as practiced today is the study of things that can be seen and touched by all. If it is concrete, tangible, and reliable, then it is likely to be a "scientific truth".

That is an oversimplification, of course. Newton's observation of the apple consistently falling down was noticing something anyone could see with their own eyes—he just put it into concrete verbal form as the law of gravity. These days, however, what we call "science" has gone far beyond the realm of the average person's perception of perceivable "reality".

The cure of mysterious diseases by new miracle drugs or the juggling of tremendous figures by a small electronic computer are far beyond understandable or explainable reality of most individuals. In fact, accepting an electronic calculator's total of a long string of figures is really more of a leap of faith than most of us would risk in theology, particularly as few of us have any better idea of how sub-miniaturized electronic micro-circuitry works than we do of the most abstruse versions of cosmology.

All we really see is that it seems to work. A vast array of gadgetry manages to get us on the moon, bring us picture stories on the TV and, in general, supports us in an infinite variety of tangible, technological ways that few of us understand entirely and most of us mainly take on faith.

Yet Uranus, the planet of clarity and precision, is given to rule all these things we don't understand. The obvious answer is, of course, that *somebody* out there discovered it, understands it, and that is why it works. The individual efforts of each of us to delineate and correlate our own existences with those of others has managed to make up a vast and complicated society which is usually understood only in limited detail by the individual, simply because there is just too much data around for any one person to comprehend.

It is the mark of the scientist, the inventor, or the innovator, to come upon new physical territory that has not yet been trod upon. Just as the religious visionary comes upon a seemingly delineated part of spiritual truth, so the scientist uncovers a highly defined niche of physical truth that had not been observed before.

The world of the inventor or the scientist is often characterized by the same style as that of the visionary. Information comes in sudden bursts. In the case of the religious observer it comes in visions or inspirations, while in the case of the scientist it happens when suddenly all the seemingly disparate facts come together and reveal a startlingly obvious truth, previously unobserved. In the case of the researcher, it may be the result of years of careful collecting and correlating without effect then, rather quickly, the evidence indicates the postulation a theory, or an unsuspected explanation made plain.

In the case of an inventor, there may be a long period of seemingly fruitless experimentation, until suddenly an idea or solution manifests. The term "inventor" simply covers those of us with a talent for dealing with materials but without the extensive training of the so-called modern "scientist". The principle is the same. The discovery of atomic energy and the discovery of the wheel were both doubtless preceded by much unsuccessful struggle with physical materials.

Often, the scientist is acclaimed as being a great "visionary" after coming upon some particularly useful piece of truth. This title is more appropriately applied to a prophet than to a technician.

Uranian vision, however, is indifferent. When something is true, physically or spiritually, it's true. Both scientist and prophet are revealing to mankind another aspect of the truth that comprises the total reality of the universe. Great thinkers in either science or religion have a deep respect for both fields, as the essential processes are the same. The truth-seeker sows his seeds, either as spiritual seeking or physical research. His rewards often come from the least expected directions, or indeed they may not come at all. Both science and religion are very unreliable in terms of results. The most arduous researcher may never come up with anything startling, while the most casual inventor may come up with an earth-shaking discovery. Similarly, the most sincere devotee may never be granted a divine vision while a wandering reprobate may find himself suddenly staring into the face of God.

If the keyword of Uranus is Truth, then the scientist's path is to uncover (or stumble upon) the physical aspects of that truth. The road of the religious leader lies in making contact with more spiritual or

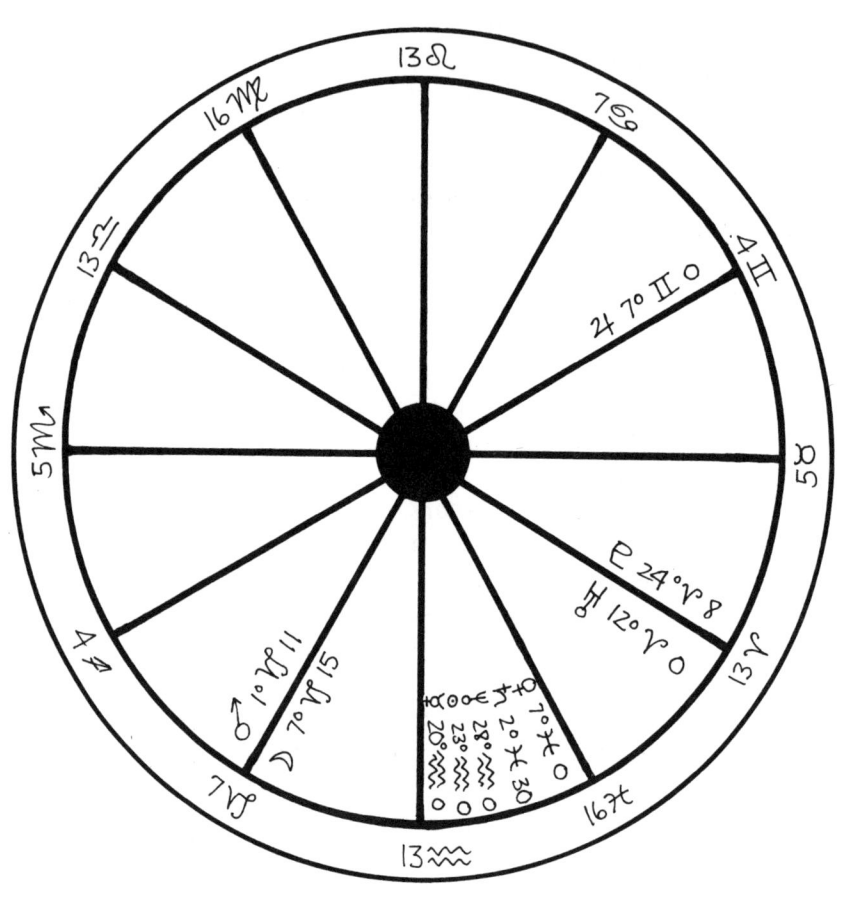

THOMAS A. EDISON

FEBRUARY 11, 1847

MILAN, OHIO

subtle truths. Looking at the birth charts of prominent scientists reveals a level of prominence of Uranus as in the charts of religious leaders. The principle is the same; only the manifestation is different:

> Einstein—Uranus ruling Ninth House, disposes Jupiter, opposition Jupiter in Third House
> Niels Bohr—Uranus opposition Moon, square Saturn
> Galileo—Uranus square the Sun
> Benjamin Franklin—Uranus opposition Mercury, square Saturn, disposes Mercury
> Bertrand Russell—Uranus ruling the Fourth House, conjunction Jupiter, opposition Saturn
> Thomas Edison—Uranus ruling Mercury, Sun, Neptune, square Moon
> Henry Ford—Uranus ruling Moon and the Third House, square Venus
> Buckminster Fuller—Uranus square Mars

Uranus represents the buildup to the sudden flash of truth and realization. In both science and religion it is characterized by conjunctions, oppositions, and squares in the natal birth charts. Hard aspects are natural to Uranus, as they bring out its native qualities with the greatest force. The greater the tension, the greater the release. The essence of vision is breakthrough, and that is seldom realized when Uranus is too easily aspected in the chart.

Whether physical or spiritual, the visionary and the innovator receive the greatest persecution of any individual in society. The tensions of Uranus set him apart, and its visions may either elevate or destroy him, but one thing is certain: the easy life of oneness with society will seldom be his lot. The person with Uranus in hard aspect in his chart will probably be very special, whether he likes it or not. Those who can sustain the tension and learn to use it may become the great scientific and spiritual leaders, while those who crumble under its harsh and unrelenting pressure become the outcasts of the world, the changelings of fortune lacking self-control.

V
Uranus, Self-understanding, and Psychology

If there is any endeavor that purports to be the combination of science and religion, it is psychology. In the nineteenth century as religion began to crumble under the assault of modern science, psychology sprang up as the scientific approach to assuaging the problems of the soul. There was a belief that a rational and analytic methodology of examining the problems of the psyche could yield more results than the supposedly irrational teachings of religion.

Of course the founders of psychology did not have quite such a large goal in mind. Freud was simply looking for ways to effectively treat certain mental disorders, not to build a substitute for religion. However, the social response to his startling new view of internal human motivations, and the remarkable results his theories brought when put into practice, was such that society began to look upon the new science of psychology as the place to look for the answers to the age-old longings of men for understanding and illumination.

Freud, in his teachings, never suggested that psychology or psycho-therapy would have any such result. He believed that you simply learned to accept the traumas of childhood and then lived with the results, like them or not. Psychopathic symptoms might be relieved by such self-awareness, but there was never any intention to provide psychic or spiritual deliverance. As far as Freud was concerned, sheer stoicism was the only way to live with your problems once you unlocked them.

Jung's theories, however, went much further toward combining science with religion. He proposed deep, mystical inner archetypes that formed a tappable subterranean lake of human consciousness, far closer to the older religious concepts of the spiritual world than perhaps Freud would have accepted. Because of his writings on astrology and more occult matters, Jung has ever since been the favorite of would-be occultists who do not wish to step too far out of line with accepted psychological ways of viewing the personality. Although Jung was a mystic at heart, he never openly espoused occultism and consistently proposed that all of the mysterious phenomena of the occult could be seen and understood rationally through scientific method and observation. In the absence of Uranian visions, the Uranian method of careful observation and notation would achieve the same goal.

Most psychologists were not as spiritually aware as Jung, however, and the very physical way of approaching the non-physical predominated, reaching its apex in behaviourism and B.F. Skinner. He believed that through sheer dry observation and environmental manipulation the personality could be directed at will. This was, and is, the most externalized way of looking at the personality. Its uses in therapy are many and often very quick and startling, reflecting the Uranian nature of the approach.

The degree of dryness (and lack of Neptunianism) with which one approaches a subject may be a reflection of how harsh and exact Uranus is within the natal chart. Jung's Uranus was closely square his Moon but a full eleven degrees from his Sun. In addition, the Moon was besieged by Neptune and Pluto, thus blunting the precise quality that Uranus might have lent.

Freud's Uranus is conjunct his Sun (within four degrees) and is the midpoint between his Sun and Mercury, thus heightening its effects. B.F. Skinner's Uranus, however, is strongest of all, being exactly square his Sun within minutes of arc.

What is common to all three, as well as to other leading thinkers in psychology, is that whatever bit of truth they have uncovered is seen clearly and precisely enough to be of effective use in therapy.

Though the approach may vary, the Uranian factor remains the same. Wilhelm Reich, far from a behaviourist and radically different from Jung or Freud, displays Uranus conjunct Saturn, square Jupiter. Dr. Benjamin Spock, whose approach to childrearing became the mainstay of Twentieth century families, also has a strong Uranus, square Mars and ruling Saturn. The more modern schools of sex therapy also show the Uranian influence. The chart of Virginia Johnson (of Masters and Johnson) has Uranus opposition Moon, ruling Sun, Mercury, and Venus.

Uranus may be said to characterize the whole process of psychotherapy, as well as the study of psychology itself. Although some forms of therapy may be drawn out over many years of treatment, such as in psychoanalysis, success or major progress in any form of treatment is usually characterized by sudden flashes of insight and abrupt behaviour pattern changes. The rhythm of Uranus is essentially stacatto, indicating a gradual build up of pressure followed by a sudden release. Similarly, most therapy involves just this kind of rhythm, and pressure is increased on the personality in

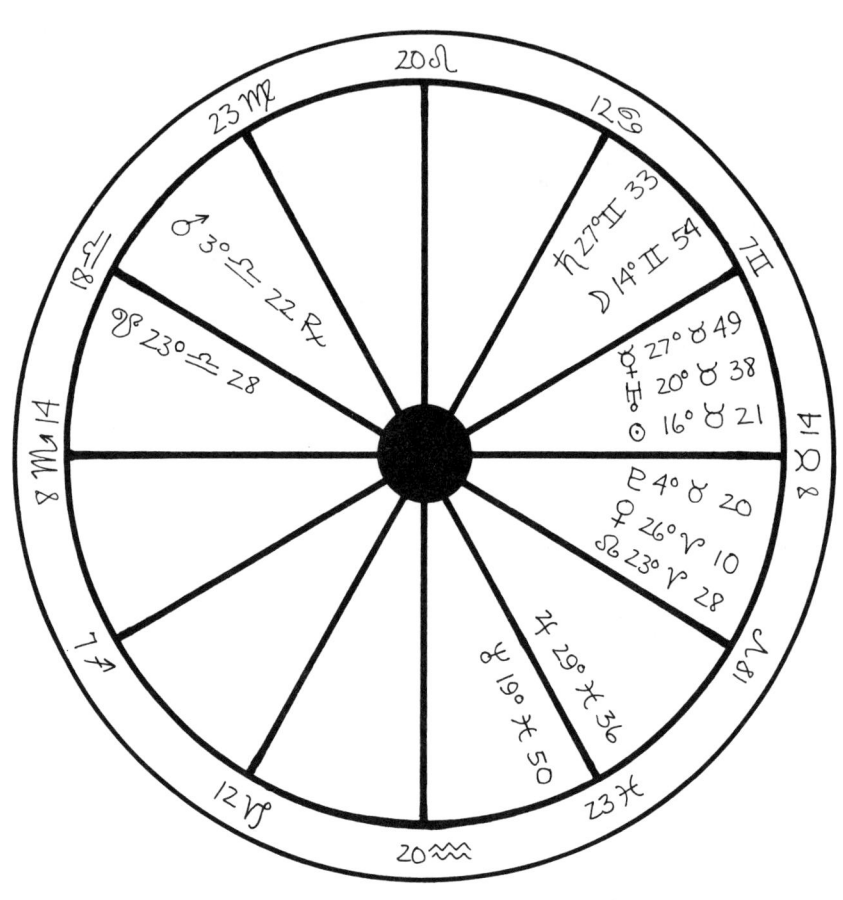

SIGMUND FREUD

MAY 6, 1856

6:30 PM

FREIGBURG, GERMANY

many ways. Among these are probing by the therapist or patient himself, peer pressure (as in group therapy), or the change of external patterns of operation (as in behaviour therapy). These pressures finally lead to a breakthrough for the patient. He gains a sudden new awareness of his psychological workings. There is also a release of internally produced tension caused by personality problems, and the external tensions caused by the therapy itself.

In the normal course of events the personality undergoes gradual changes in self-awareness over periods of many years, particularly as described by the standard astrological cycles we all must share in common. However, through various causes these normal changes may be blocked, causing the personality to become out of tune with the body, other personalities surrounding it, and with itself. This state may be described as neurosis or psychosis. Sometimes this is caused by improper chemical changes that are quite physical, but more often caused by faulty adaptation of thoughts and behaviour patterns. At this time, in therapy, pressure is brought to bear to explore and break through such blocks, thus freeing the path of the psyche to again become in tune with itself and the environment.

In earlier societies, such Uranian insights and personality changes were sometimes brought about by religious conversion. More often than not personality "diseases" were simply left untreated until the internal dams burst of their own accord in violence and psychosis. If this did not occur, the blocked individuals continued to lead hampered lives, or withdrew entirely and were rejected by society.

Thus, thanks to the blending of the Uranian principles of science and religion, a new manifestation of Uranus has arisen that has helped to ease the pressures of society and has made life worth living again for millions. Its only danger, in a true Uranian sense, is that it can at times become too self-obsessed and exclusive. Uranus tends to look upon itself, even at the lowest levels, as the whole and exclusive truth of the matter. Psychology is not the only or total description of the human being, and there is danger in assuming it is so, as many enthusiastic therapists do. Janov, for instance, proposes primal therapy as "*the* cure for neurosis", not *a* cure. Such Uranian extremism serves only to damage the structure of a great and necessary new field of human endeavor.

VI
Uranus and Creativity

If there is any area in which Uranus and its effects become ambiguous it is in the arts. Traditionally, the creative arts are associated with Neptune. Yet an inspection of any large sample of nativities of notable achievers in music, painting, writing, and films, will turn up a startling display of Uranus.

Where in science, religion, and psychology the Uranus factor seems almost a necessity, in the arts its presence becomes more a matter of style. In the field of acting, for instance, someone who never plays Uranian roles can get on quite well without a prominent Uranus (Greta Garbo, for example). Someone who relies upon brusqueness for his style will require a very strong Uranus (such as John Wayne, with Mars conjunct Uranus opposite Jupiter conjunct Neptune, the perfect combination picture of the two-fisted, hard-fighting idealist).

The essence of the great actor, however, is to be able to play many different roles. In general, therefore, you find the best all-around actors to have well-balanced charts including a strong, but not overpowering Uranus. An overly strong Uranus (or any other planet) will tend to stereotype the artist, as in the

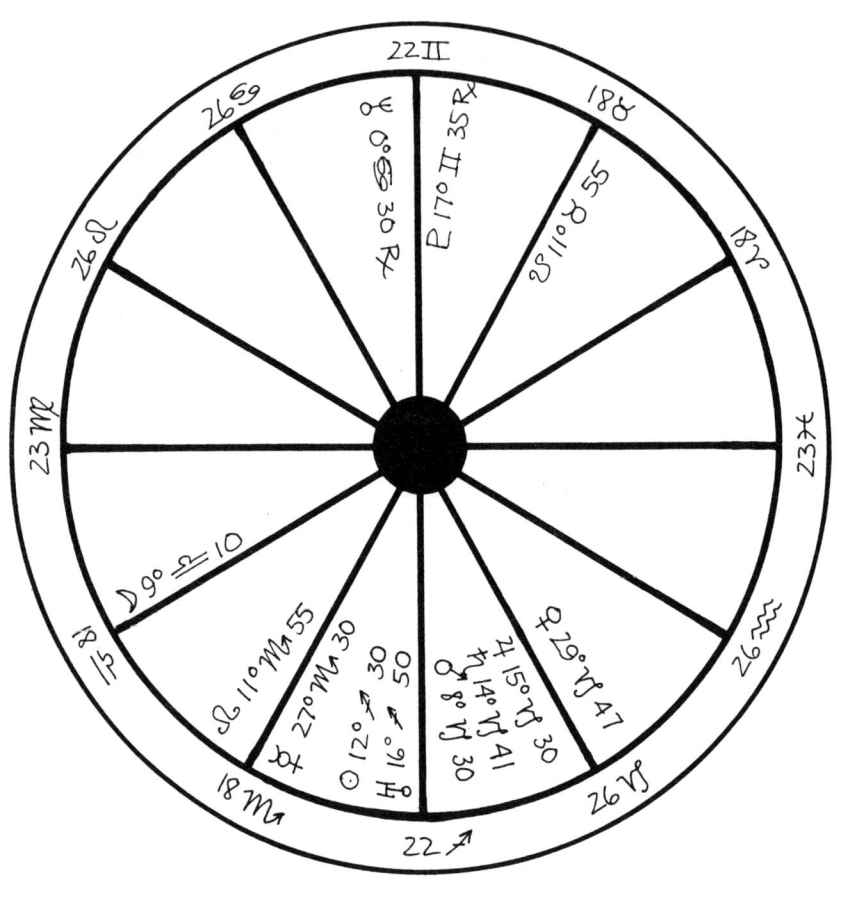

WALT DISNEY

DECEMBER 5, 1901

12:30 AM

CHICAGO, ILLINOIS

case of Errol Flynn. Here Uranus is on the ascendent opposite Venus with Aquarius intercepted in the first house. In addition, Mars opposes Jupiter, which completes the image of the daring adventurer, which Flynn was both in real life and in the movies.

Uranus is much easier to spot behind the camera. Because a good film director must shape his creativity with a detailed knowledge of the technology of his art, Uranus manifests in specific physical truths.

D.W. Griffith, who was responsible for inventing most of the common camera and directorial methods still used today, had Uranus conjunct the Moon, opposing Saturn, and ruling Sun, Mercury and Saturn. Certainly Uranus here represents the technical innovator as well as the creative artist.

Similarly, Walt Disney, another great film innovator, displays the Sun conjunct Uranus opposite Pluto and MC. His genius ran the gamut from creating the antics of Mickey Mouse to running a multi-million dollar entertainment empire.

More recent directors such as Alfred Hitchcock (Uranus square Mercury) display similar aspects.

In music, the presence or need for Uranus varies widely. Among modern pop singers, for instance, Uranus aspects are conspicuously absent. Among pop songwriters as well, the planet seems to take the back seat.

Among country-western singers, of all people, the story is just the opposite. It seems that in order to be a successful country artist, a strong Uranus is an absolute necessity. At least so in the charts of Hank Williams, Jimmie Rodgers, Merle Haggard, Rosemary Clooney, Johnny Cash, Patsy Cline, Pat Boone, Elvis Presley, Glen Campbell, and Dinah Shore as a brief sampling would seem to indicate. This appears strange, considering the Neptunian sentimental nature of the musical style.

Country music does have two very Uranian qualities, however. One is a generally strict form and structure, although there is no prominence of Uranus among blues singers, where the musical structure is even tighter. The other arises from the culture itself. The songs spring from a culture steeped in sexual repression, and the tension and violence that result from it. Both of these are very Uranian in nature. The ability to express these tensions is paramount in becoming a successful country singing star, which probably explains the overwhelmingly strong Uranus picture in that field.

Uranus is very prominent among painters. Few painters become famous (or have in recent history) unless their work is new, original, and challenges the older styles of painting, all Uranian traits. Painters who were either unaware of tradition (like Grandma Moses) or were mainstream illustrators (like Toulouse Lautrec) often did not have a prominent Uranus in their charts. This may be a further insight into why Uranus's influence is absent in pop and folk singers' charts. In both cases, there is no tension with society. The folk artist or singer rolls along his merry way, not aware of any need to conform, while the pop artist is the ultimate conformist stylistically or he cannot succeed. It is only the person aware of artistic conformity—and breaking it—that is characterized by Uranus.

Classical music, however, presents a wide spectrum of variations as far as Uranus is concerned. Uranus runs fairly strong among classical composers, which would seem logical considering the vision and technical knowledge it takes to write the inspired and complex pieces necessary in the field.

But certain composers just lack it where they seemingly should not: Bach, for instance, whose clear vision and technical virtuosity surpassed all others; Beethoven, a revolutionary in the music of his day as well as a tremendously capable technician. Beethoven's Uranus is involved in a grand earth trine, hardly likely to produce the tense, explosive character that marks his music and his personality as well. Bach's strongest Uranus aspects are a wide trine to Saturn and a sextile to Neptune. Certainly he was a more peaceful sort than Beethoven, but theory would have one expect a more prominent Uranus.

But perhaps it is well that all astrology does not work for us as predicted, or we would all be pre-programmed robots. When we speak of the position of Uranus in relation to the arts, or in any other field, we must remember that we speak in general terms, and specifics do not always follow so easily. Beyond the supposed temperaments handed down to us at birth, we are all individuals capable of almost anything, regardless of our nativities. The very existence of astrological twins (indeed, most of us have one somewhere) is proof that the natal chart is only a filter between the soul and individual circumstance. Change the soul or the circumstance and everything changes, even though the aspects in the chart remain.

So also we must consider artistic creativity to be the expression of the soul, molded somewhat by the natal chart and channelled by external opportunity, but pure in itself. The more pure and complete the expression, the greater and more moving the work of art.

Art is not *motivated* by any planet in the nativity. Its expression may be triggered by transits to various points in a chart, but the motivating force comes from within or, if you will, above. Neptunian inspiration and Uranian realization both channel the creative urge, but if such an urge can be attached to any one point in the chart it would be to the Sun itself, the center of creative life. Uranus, Neptune, and the other planets in the chart are the tools wherewith that urge may be shaped and sculptured, but the prime life force that is creativity remains a thing apart, the deity or muse behind the many forms of man's artistic expression.

VII
Uranus, Travel, and Adventure

As Uranus represents the new and the challenging, it is associated with travellers and adventurers of every sort: those who traverse the world and those who invent new horizons and adventures in their minds. Uranus does not represent travel for leisure or entertainment, but rather for the sake of learning new things and determining new truths about the world.

This does not necessarily mean that one must discover new truths for mankind in order to qualify—new truths for oneself is all that is needed. Errol Flynn, who certainly had the chart and the personality of an adventurer, did not add to the knowledge of any unexplored territory for mankind in general. His Uranian contribution was in the area of film art, making adventure come alive for a generation of armchair pirates and swashbucklers. He travelled all around the world both before and after he became an actor, gratifying an insatiable lust for the new, the unusual, and the adventurous. He has been criticized for not being a great actor, yet the very excitement of the personal life he led is put into the characters he plays and his films are the greatest of their kind, which no one will deny. His life, and those of the characters he portrays, was not one of great depth or intellectual might. It was one of continuous excitement, danger, and variety that would challenge and defeat men with less fire and courage. These are the Uranian, and primarily masculine, qualities of the adventurer.

One need not be a man to display this kind of spark and courage. Aviatrix Amelia Earhart is a prime example of an adventurer on the distaff side. Her chart is featured by Uranus conjunct Saturn square the Sun and ruling the MC. Indeed, she was, like Flynn, a professional adventurer (Uranus MC)—not a movie lot explorer, either, but the real thing. Considering Uranus' given rulership of the air and airplanes as well, she couldn't have picked a better vehicle in which to do her exploring! Unfortunately, like so many other great explorers and adventurers, her profession was the cause of her death.

The urge to explore, to open up new territories, is native to the United States as a country. Uranus is prominent in the Declaration of Independence chart no matter what time is used, and the style of the country has been very Uranian for all of its 200-odd years. Daniel Boone, one of the first explorers of the Western wilderness was born with Uranus opposition Jupiter. George Washington, an explorer and surveyor in his youth, also had the necessary prominent Uranus, but his expression of that planet was to show greater and more significant variety as he matured. Uranus is important in Theodore Roosevelt's chart. He travelled and hunted for fun, which shows with Uranus quincunx the Fifth house Sun, ruling the Ninth, but his style and personality were always that of the brusque and bristling cavalryman and explorer that he was. Uranus is also to be found prominently in charts like that of astronaut John Glen (Uranus square Venus, opposition Jupiter).

Exploration need not be on the physical plane to be characterized by a strong Uranus. When the mind travels to new lands and meets new challenges the same

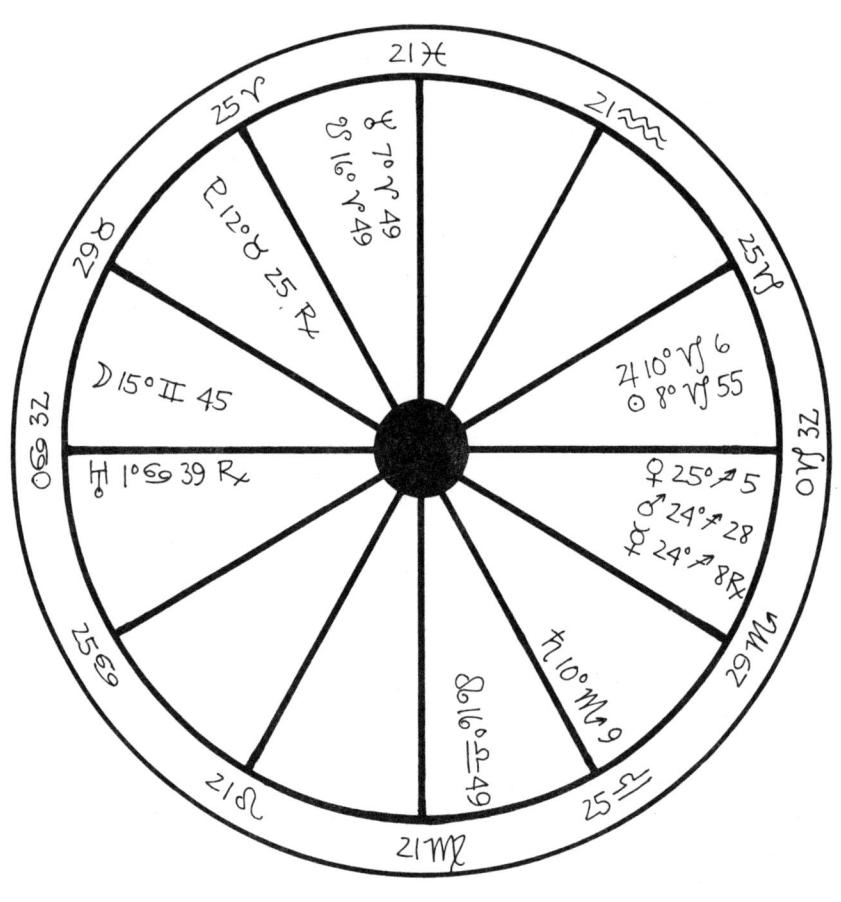

RUDYARD KIPLING

DECEMBER 30, 1865

4:53 PM

BOMBAY, INDIA

forces are in play. Great adventure writers are exemplary of this. Jules Verne (Uranus in the Ninth ruling Sun, MC, and Mercury) and Edgar Rice Burroughs (Uranus in the same degree as Verne's square Venus) expressed adventure fantasies in their writings and millions of readers lived them. (This may seem a very Neptunian expression of Uranus, but then if that planet were manifested physically all the time there would be nothing but chaos and confusion. Better some of it should remain upon the mental plane . . .)

Rudyard Kipling is most representative of the Uranian adventure writer. His topics included other Uranian subjects such as war and the visionary occult, and his chart clearly indicates that, with Uranus conjunction Ascendant, opposing the Sun and ruling the Ninth house.

Travels are not all made over the surface of the earth or even in the pages of a novel; some are made solely in the mind itself. It is particularly interesting that the different varieties of psychedelic experiences that occur under the influence of psychogenic drugs such as LSD are called "trips." Indeed, this colloquialism is appropriate. The individual who has ingested such a drug has journeyed as far from the ordinary everyday reality as does a mountain climber or a polar explorer, perhaps further.

These "interior" journeys seem to offer as much variety, interest, and hazard as any explorer might know, if the casualty rates of both are taken into account. Psychedelic explorers display a prominent Uranus, as in the example of Timothy Leary, who has Uranus conjunction Moon, square Sun, ruling Moon. But here we have gone almost full circle back to religion. Is not the revelation and truth sought by the inner explorer essentially of the same quality as that the physical explorer seeks?

This begins to be very Neptunian ground when drugs are involved. Yet the psychedelics display, in their effects, as much Uranus as Neptune. Are the blindingly sudden visionary effects of dimethyltriptamine any less real than the waking visions of the prophets? Are the sometimes clearly delineated places to which the mind journeys on an LSD, mescaline, or psillocybin "trip" as real as the far-off peaks the explorer mounts?

The results are simply not in yet. The Uranian exploration into areas previously given over to Neptune has just begun in the last few decades. Perhaps, when all is said and done, Uranus will shed its harsh light upon all that is now in Neptunian shadows. We may realize, as Buddha proclaimed, that the harshest of Uranian realities is simply Maya, pure illusion.

But that time has not yet arrived. The universe still bears seemingly endless exploration both internal and external, and it is a fairly reliable statement to say that the children of Uranus will be the ones to take up the task.

VIII
Uranus and Humor

Theorists have long debated the nature of humor. What is it that makes people laugh? What is the key to a good joke? Does humor simply mask reality with a sugar coating, or does it lend deeper perception into the nature of the human predicament? If anyone knew the answer to such questions he could make a fortune in the comedy business, but so far there have been no takers. Many people have done well in comedy on stage, screen, and print, but each has a different version of what makes people laugh. For each of these some people will laugh and some will not. The best of comedians has bombed many a time.

Some psychologists say laughter is a defensive reaction against fear. This may be an overstatement, but it has its merits. I remember a roomful of people clutching their sides at a joke about how a stupid woman tried to put her sick dog away by roasting it in the oven alive. With the narrator's description of each agonized jump of the ailing dog the spectators laughed louder. What they were hearing was hardly a joke if they had thought it was really true (of course it was presented as if it were true). It was a horror story of cruelty and pain. Admittedly my own reaction to it, thanks partially to the narrator's style, was laughter as well. Yet it caused another person in the room, who was not so given to comic horror, to break down in tears.

If what makes one person laugh makes another cry, then what, indeed, is humor?

Certainly the average limerick or shaggy dog story does not strike fear into the heart of the listener. The average comedian's formula for drawing a laugh is surprise. If the punch line of a story is surprising enough then laughter is sure to follow. This may be a better definition of humor. If this is a correct description, then Uranus is the planet that rules humor. The sudden, the unexpected, the quick turnaround and twist are all native to the qualities of Uranus.

Within all of us there lingers some "feeling" closely akin to fear. We feel safe and happy when there are no surprises. When something new or unexpected comes upon the horizon we feel we must first investigate it to see that it brings no harm before it is permitted to enter our lives unchallenged. Surprise and alarm can be no fun at all.

Yet humor is the very essence of what we consider fun and enjoyment. "Laughter is the best medicine" we say, and so it truly is. A good laugh goes a long way to cure our problems, at least those that arise from within ourselves. How can something that induces surprise and possibly even fear make us happy and content? It seems a total contradiction.

Perhaps it is not. We find nothing funny about a pie being thrown in our own faces. It would be most disturbing and unpleasant to say the least. Yet we still laugh when we see that age-old routine presented on the screen, where it is happening to others. That may be the key.

Certainly no man was more aware of this than Mack Sennet, the founder of slapstick comedy in the movies. The endless trials and often painful tribulations of the Keystone Cops were a boundless source of laughs for the silent screen audiences. Sennet developed all the basic principles of sight-gag comedy, to be later embellished upon by his own protege Charlie Chaplin. Sennet had Uranus conjunct Moon, square Saturn, giving him a natural ability to sense surprise oriented comedy, mixed with an ability to painstakingly set up the necessary actors and props to achieve it. Chaplin, with the more immediate sense of an actor, had Uranus conjunct his Mercury. He wrote all his own material, and he even wrote the music for his later sound movie scores. Yet both men were harsh and demanding directors, not satisfied with a scene until it had reached its pinnacle of truth and perfection.

Other classic comedians were more relaxed. W.C. Fields with his Moon conjunct Uranus was one of the most relaxed. Yet he was always there with that sudden wry reaction that could set an audience reeling.

Comedy has many different styles, often marked by the Uranus aspects of the performer. A classic set of opposites was Jack Benny and George Burns, affectionate adversaries through decades of stage, screen, radio, and television humor. Benny's jokes were seldom directed with malice toward anyone with his Uranus opposite Jupiter in the 5th, ruling Sun and Venus. His obvious generosity was mocked by his pretended stinginess. His adversary George Burn's style was marked by Saturn conjunct Uranus, square Mercury and ruling Mercury. His jokes were always laconic and played close to the chest.

Modern comedians inject social and psychological satire into their work. A good example is master comic filmmaker Woody Allen. His Uranus quincunx Mercury ruling Moon lends a restless, relentless and stinging clarity to his humor. Like so many modern comics, his jokes often strike a little too close to home for comfort.

The fine line between the surprises and difficulties of others, and these same problems when

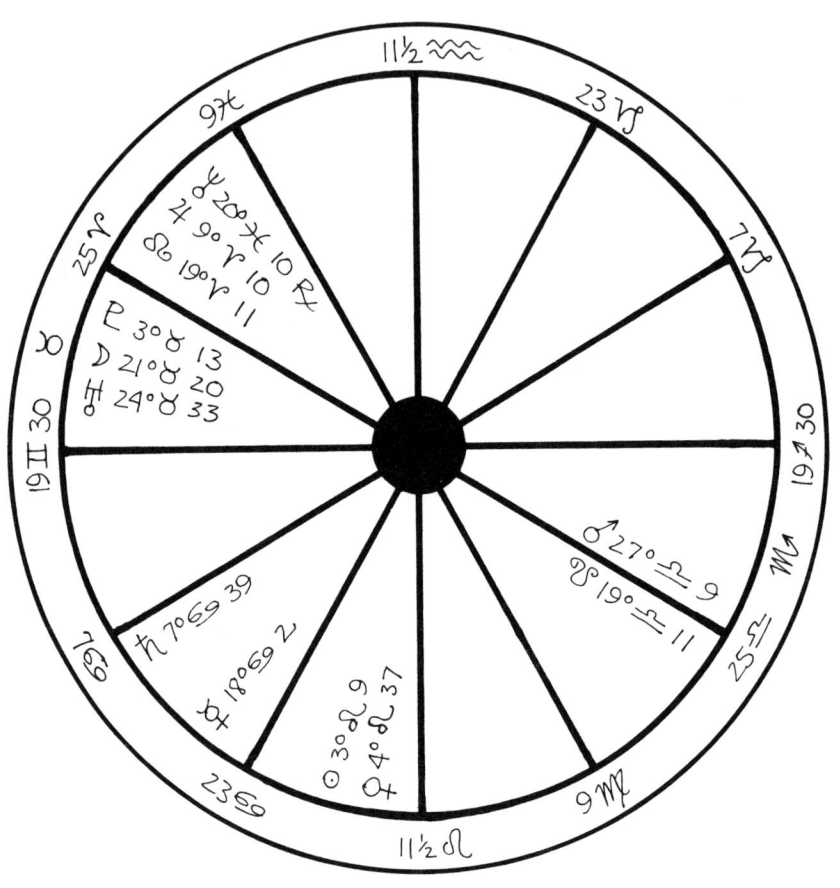

GEORGE BERNARD SHAW

JULY 26, 1856

12:40 AM

DUBLIN, IRELAND

they happen to ourselves, may be the fine line that the best of comedy straddles. Too much obviously contrived slapstick tends to be boring and miss the desired effect. The viewer cannot empathize with the subject at all and therefore loses interest. Too close of an identification with the comic actor also loses the effect of humor and tends to become tragedy and pathos.

The best of comedy has the viewer still safely on his side of the screen, but just dangerously close enough to titillate. This is the element of Uranus contained, held at arm's length where it stimulates but does not threaten too much. This appears to be the most successful and popular kind of humor.

One kind of humor goes over the line intentionally. That is so-called "black humor". Jonathan Swift is probably one of the earliest of modern Western examples. Although the style dates back to early Greek comics. Swift's "Modest Proposal" about raising Irish babies to eat makes the blood run cold, yet it was meant as a kind of socially stinging comedy.

The classic 20th-century master of the art was George Bernard Shaw. His humor was often wry and sophisticated but, equally as often, so biting as to be offensive to his native Britons. His ability to make humor dark and painful may come from his Uranus conjunct Moon in the 12th, ruling the 10th, as he certainly made his reputation with it.

Whatever the brand of joke, it is Uranus that gives it the punch, whether mixed with Martian qualities, as in slapstick, or those of Mercury, as in more intellectual and sophisticated humor. It is most popular when it is kept fairly safe for the viewer. But Uranus cannot always be kept safe, that is not its nature. Thus humor covers all of the ground from expectant father jokes to the grave . . .

IX
Uranus and the Bizarre

What is it that makes people queue up in lines to file past a deformed dwarf, gawking at his truncated body and twisted limbs? What makes people almost involuntarily stare at some knotted freak of nature such as the two-headed calf, the bearded lady, the pin-headed man?

Whatever it is has filled the pockets of circus producers since Classical times. There is something strangely compelling about the sight of the bizarre in human form or, for that matter, in any other form. The feeling is one in the pit of the stomach not unlike the feeling one gets when approaching an accident on the highway: is anyone hurt, are there any bleeding bodies lying amidst the shards of metal and glass on the roadway? Not pleasant thoughts, but attractive enough in a perverse way to cause bottlenecks of rubbernecking motorists filing past fatal accidents to tie up rush hour traffic.

In films the love of the bizarre is allowed to run as rampant as a Hollywood makeup man's talents will permit: the mummy, Frankenstein's monster, crazed zombies and werewolves. There is nearly as great a demand for flim-flam fright as there is for the tickling surprises that comedy brings.

Both comedy and the fascination and horror of the bizarre are characteristics of the more frightening aspects of Uranus presented in a way that may titillate but not actually harm an audience. The scary aspect of an advancing werewolf may set the adrenalin flowing in the veins of a television viewer with very positive results, but to turn around and suddenly behold the menacing gun barrel of a midnight burglar wouldn't be half so enjoyable. This is not dissimilar to the Uranian enjoyments of the armchair adventurer. Even to the real-life adventurer, tight situations seem more enjoyable in retrospect than when they occurred. Still there is the ever-present aspect of Uranian truth in all these elements. A plastic image of a deformed human would not be interesting in the least, it is the knowledge that the pitiable creature is *real* that provides the thrill.

In the last few decades certain tabloid newspapers which purport to show the most grotesque and grisly scenes of bizarre sex-murders and the most unspeakably awful crimes have achieved great popularity. These stem from a long popular tradition of enjoyment of the startlingly grisly and bizarre. The oldest folk ballads, dating to the Middle Ages, are about the bloodiest of murders, mixed with incest, rape, and sodomy.

The sales of these sheets is maintained by a belief on the part of the readers that what they are reading really happened. This is not the case as these tabloids concoct most of their stories and doctor pictures to fit. But any reader that subsequently finds this out is totally denied any future thrill from reading it. It is not merely the gory and fanciful imagination of the paper's creators that interests him, that is secondary. It is that he really believes that what he is reading is *actual fact*.

The greatest purveyor of this kind of journalism was Robert Harrison, whose magazine *Confidential* broke all newsstand sales records with a five million monthly circulation in the middle to late 1950's. The magazine purported to tell every kind of lurid fact about the popular public figures of the day, emphasizing revelations of a sexual nature. The magazine, to give it credit, carefully documented all its stories and was, indeed, giving the most lucid facts, if a bit editorially slanted. But Hollywood couldn't stand the heat, and eventually a California State criminal libel suit forced Harrison to sell the magazine, despite the fact that his facts were generally impeccably researched. Some truths are just too difficult for some people to swallow.

What is particularly ironic, however, is that the same publisher subsequently opened a tabloid scandal sheet in which the stories were almost entirely made-up, and far gorier and more pornographic than his previous magazine had dared venture. Yet there was never a challenge to his new enterprise, because it had only the semblance of that Uranus-truth and could therefore only titillate the reader.

Harrison shares the same Uranus aspect with another great American profiteer of the bizarre, P.T. Barnum. Both have Uranus square Venus, and in Harrison's case it rules the 5th house where the brunt of his projects lay. The stimulus of the bizarre served to produce money in both cases, one in the publishing world, the other in the more down-to-earth world of side-show freaks and carnivals.

The ability to couch the truth in ways that startle or make one uneasy is a Uranian trait that penetrates many areas. Franz Kafka's many unsettling stories provoke social commentary and are indeed surrealistic, perhaps marked by Uranus square his Venus and Mercury.

Oddly enough, Rube Goldberg, born the next day of the same year shares a similarly Uranian quality, yet his is in humor. The sheer workability of his complicated means to an end is both startling and very funny. In art we find similar trends, most typified by Salvadore Dali, with Uranus square his Moon and MC. No less unsettling are the etchings of Maurice Escher with his involved spatial illusions, represented in his chart by Uranus opposing Moon and Mercury. Actors who play bizarre parts are also characterized by Uranus, which gives them the ability to do so, as in the cases of Boris Karloff (Venus square Uranus), Lon Chaney (Uranus opposition Mars and Mercury), and Vincent Price (Uranus square Jupiter).

In the Greek myth, Uranus was the father of a host of strange and monstrous offspring which caused great strife and trouble to the world and eventually even murdered the god that spawned them. In similar fashion the planet Uranus represents, in one of its dark sides, the frightening and unnatural manifestations of existence. These are undeniably real and, although seemingly harmless in this modern day, still seem to threaten the existence of the more ordinary of us.

But here we must learn again that Uranus is never ordinary, but is always true. Such disturbing truths give nightmares to the best of us in sleep, but while we are safely awake may provide a queer and compelling fascination: seed of just such monstrosities, inherited from our proto-Father Uranus.

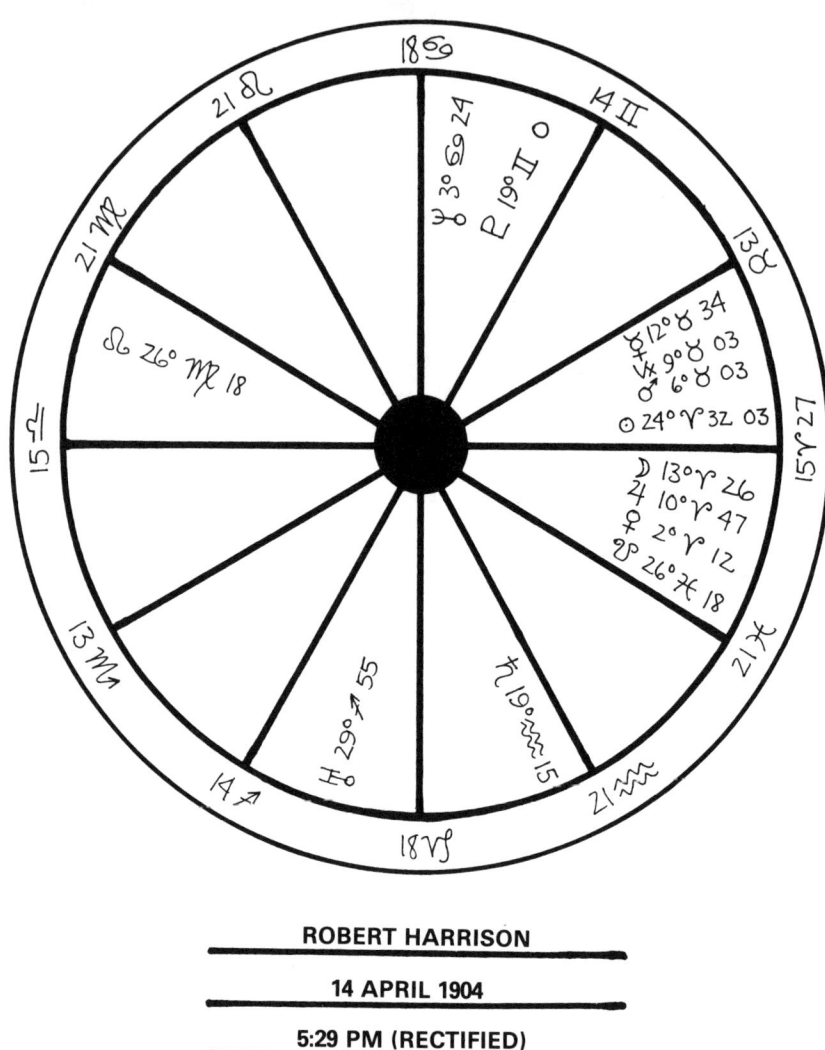

ROBERT HARRISON

14 APRIL 1904

5:29 PM (RECTIFIED)

NEW YORK, N.Y.

ROBERT HARRISON

Robert Harrison is a landmark figure in American attitudes toward the personally and sexually bizarre. Starting his publishing career in the early 1940's with the successful pin-up magazine *Beauty Parade* he nearly singlehandedly mined the American male's latent fetishism with pictorials of sultry models decked out in spiked heels, fishnet stockings, jungle dress, whips, and chains in such top-selling girlie mags as *Flirt, Whisper, Titter, Wink,* and *Eyeful.* The key to their success was a mixture of the usual pin-up appeal with an editorial slant toward the kinky and bizarre.

URANUS AND THE BIZARRE

Harrison's greatest achievement, however, was *Confidential* magazine, the ultimate scandal sheet. This magazine dug out the most shocking and titillating details of the personal lives of national celebrities, researched each story with utmost care so as to be able to prove in court (in event of a libel suit) that each strange story was, in fact, the absolute truth. After attaining the highest magazine circulation in history at the time (5 million), Harrison's nose for sensitive Hollywood news landed him a criminal slander indictment from the State of California and he was forced, under heavy political pressure, to sell his company. In the sexually repressed climate of the 1950's, his kind of truth, phrased in direct colloquial style, loaded with hard-hitting innuendos, was more than the American establishment with its closet full of skeletons could handle.

Retreating from the careful documentation of *Confidential*, he launched into a tabloid newspaper called the *Inside News*, on whose pages every grisly and bizarre sex, dope, perversion, and murder fantasy came true. This format was a great success throughout the 1960's and was copied by dozens of other papers. The stories, on the whole, were safely made up, accompanied by retouched pictures that allowed writer and reader alike to experience their most seemingly perverse fantasies as if they were as real as the front page stories of the *New York Times*. Although the tabloid format waned in the 1970's, and Harrison's fortunes with it, the need of readers to live out their most extreme fantasies in print found a new home in the slick men's monthlies and sex advice magazines that supplanted it.

Harrison's chart is an exceedingly fortunate one for the area he excelled in. A grand air trine with Saturn next to the 5th cusp and Pluto in the 9th mutually aspected by the ascendant describes his extraordinary fortune under the most threatening of circumstances both in and out of publishing.

Uranus opposition Neptune T-square Venus with Uranus ruling the 5th aptly describes the subject matter he chose, and the Jupiter-Moon conjunction and Sun in Aries together with a Mars-Mercury-Vertex conjunction in Taurus demonstrate the aggressiveness and determination with which he pursued his career.

Uranus is most certainly the key to the chart, posited in the 3rd (publishing) squaring Venus and ruling the 5th at whose cusp sits Saturn (socially repressed sexual drives). Oddly enough, Harrison avowedly disapproves of *true* pornography, whatever that is. Truly the horoscope of the sexual P.T. Barnum of our age.

X
Uranus and Sexual Perversion

Even a brief glimpse at the legend of the god Uranus (see Chapter II) will reveal an immediate and logical connection with sexual perversion. Uranus married his mother and spawned deformed monsters who finally castrated him and chopped him into little pieces. The breaking of sexual taboos and the violence that is often associated with it are integral parts of the Uranus myth and necessarily part of its astrological symbolism.

If we are to understand Uranus' nature and influence in regard to sexual perversion, however, we must first come to some sort of agreement concerning what perversion really is. This in itself is no easy task, as the social avoidance of the subject has prevented it from being too clearly defined.

The dictionary, for instance, is of little help. It simply defines perversion as deviation, and vice versa. A second definition gives it as being that which is morally wrong, which might include anything from murder to taking bribes.

Laws concerning perversion are equally vague, and vary in their definitions from state to state, almost always skirting the issues and using such precision language as "crimes against nature", "abominations before God", and the like. When actual descriptive terms are used, it is usually that catch-all "sodomy", which can cover anything imaginable depending upon what state of mind you are in. This generic term harkens back to the crimes of the Biblical people of Sodom and Gomorrah, who presumably were perverted and thus were destroyed by the Lord. That is a large presumption, however, as the Bible never mentions exactly what their crimes were, except to generalize that they were a bad lot indeed and grieved the Lord sorely. Why Sodom, and not Gomorrah, was given credit for sexual perversions is anybody's guess, but it is of interest to note that the ancient Hebrew meaning of the word "sodom" connotes "burning", which is what happened to that city and what may well describe Uranus' role in alleged sexual perversions.

Anyone investigating the history of sexual mores and attitudes will find that there is practically no variety of sex that is not considered perverse by someone. Most cultures permit sexual intercourse after marriage in the so-called "missionary position" at certain times of the month—St. Paul suggests that even this is a step in the wrong direction, but grudgingly admits that it is "better to marry than to burn". There have, indeed, been sects throughout history which have banned all forms of sex as perversity, but they all, predictably, died off for want of propagation.

What, then, is perversion?

Perhaps, in an attempt to defuse its moral connotations, it could simply be called sexual expression that is not in the cultural mainstream, unusual sexuality. Whether it is morally wrong to be unusual in any context, sexual or otherwise, is a much-debated point in most societies but does not merit coverage here, as that is not our subject. Such moral condemnation of unusual sexual expression does, however, sharpen and perhaps define the Uranus qualities inherent therein.

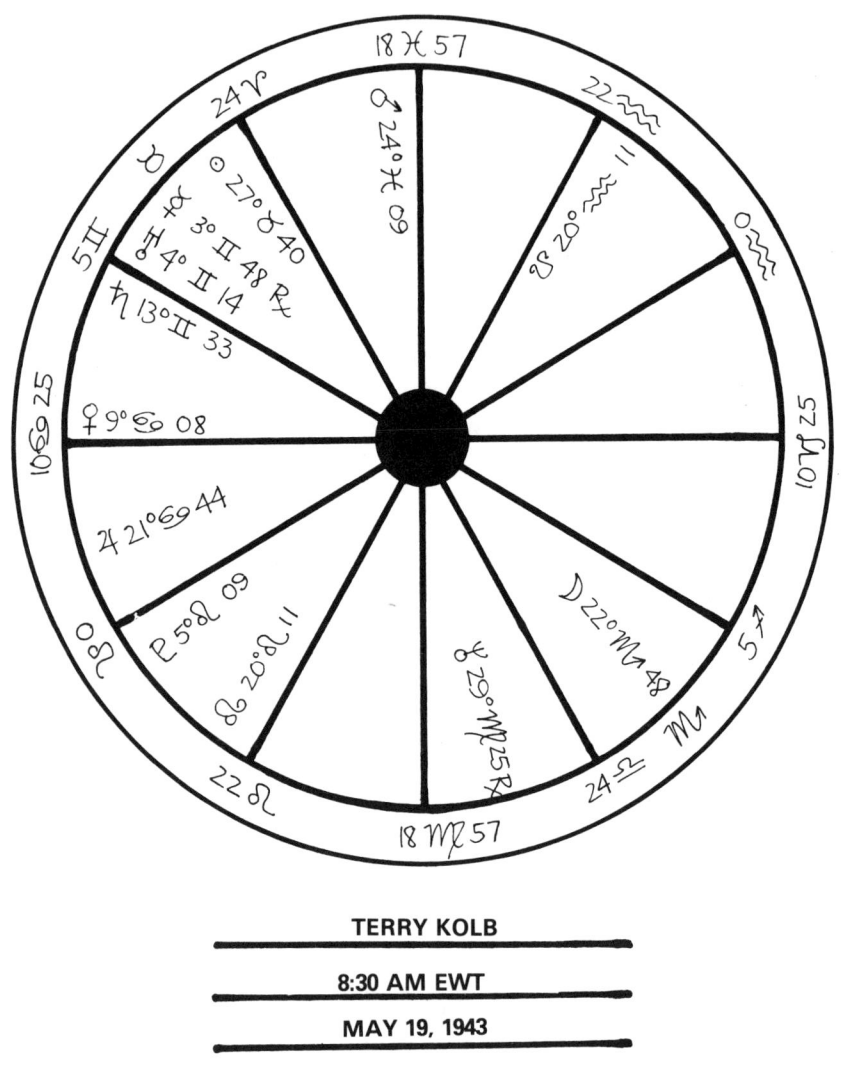

TERRY KOLB

8:30 AM EWT

MAY 19, 1943

NEW YORK, N.Y.

TERRY KOLB

Terry Kolb is one of the founders of the sado-masochists' liberation group the Eulenspiegel Society, formed in the ealy 1970's in New York City.

She is a masochist in the truest sense of the word, and is a fairly extreme example, as most people with this inclination do not go into it as deeply as she has, which has included actual physical injury upon occasions.

Uranus is conjunct Mercury at the cusp of the 12th (secret bizarre fantasies of subjugation) re-emphasized sexually by the Sun-Moon opposition along the 5th-11th houses. A grand water trine involving Moon-Jupiter-Mars with Mars near the MC allows these inclinations to flow into reality and, indeed earn a reputation for it (Mars-MC) by writing on the subject (Mercury-Uranus) in national magazines as well as helping establish an organization for better understanding of S/M.

Of interest is that, unlike others of her persuasion, she dropped out of the S/M liberation movement in favor of finding total passivity and domination, which she treats in her writings as almost a religious experience similar to martyrdom. This may be attributable to Mars (despite its elevation) in Pisces ruled and debilitated by an opposing Neptune, as well as Venus and ascendant being midpoint Saturn/Pluto (realized desire for total personal erasure). Here, the total S/M Uranus experience must eventually be death.

Traditionally, hard aspects of Uranus to the lights and inner and middle planets are supposed to be related to sexual perversion in the natal chart. In personal experience this seems to be so but, so far, statistical studies have failed to systematically link Uranus or any other planet with sexual oddities. Much of the research in this area has been extremely muddled by antiquated or self-serving definitions of what perversion really is, so negative results might well be expected.

Specific charts aside, everyone will pretty much agree that Uranus relates to the odd and unusual and thus to out-of-the-ordinary sexual practices like the diverse descriptions of sodomy, homosexuality, sado-masochism, fetishes, and the like.

If the prospect of normal sex heightens our arousal and sharpens our senses, then the prospect (not to mention the commission) of varieties of sexual expression that are off-beat and highly forbidden must necessarily produce the harshest and sharpest sexual tension. The Neptunian qualities of romance and accepted love and affection are stripped away and the primal, self-immolating sex drive takes over the spotlight. Uranus-characterized forms of sex indeed do partake of that definition of "burning" associated with Sodom and alluded to by St. Paul.

All forms of socially-condemned sex are perforce very high-intensity in nature, first because they are risky and thus produce nervous tension, and second because an individual must be very forcibly driven internally to engage in such avowedly anti-social acts. Social condemnation has always been a heightening factor in even so-called "normal" love affairs—where would the romance have been if Romeo and Juliet's families had set them up with a big church wedding? Certainly elopement is far more exciting than courting your lover in your parents' living room.

In the case of the more unusual forms of sexual expression, this intensity is heightened to an even greater degree and is thus characterized by Uranus, whose harshness and intensity is the greatest of all the planets.

In recent years, the increasing acceptance of other than middle-of-the-road sexuality has had an interesting effect on its Uranus qualities. This may be seen particularly in magazines, books, and movies that attempt to use sexual themes to arouse and titillate the audience, sort of a voyeuristic Uranus experience like the armchair explorer gets from adventure novels. It used to be quite sufficient to simply use the prospect of normal sex by showing pictures or descriptions of women and/or men in various states of undress. Of late, however, as "kinky" or bizarre sex has become accepted and indeed in vogue, the media is forced to depict more and more unusual and strange sex practices in order to achieve that titillation which the customer wants. Endless new variations are presented to provide new thrills and each in turn is dulled and defused of its excitement by overexposure and acceptance and is replaced by something even more peculiar with the hope of again achieving the desired Uranus effect.

Certainly the legal and moral defusing of many harmless but so-called "perverted" forms of sex-play is salutary. It is a mark of an increasingly enlightened society that it can at last rationally view what truly is and is not harmful in that area and reshape its opinions and laws accordingly. Hopefully, the antiquated laws against sexual expression (for its own sake rather than for the sake of procreation) are pretty well doomed. Sex has become something we may understand and enjoy rather than furtively and fearfully participate in.

Many ask of recent sexual liberation: when you have laid the whole subject bare, where's the thrill? When everything's okay and aboveboard, who will be interested anymore? Admittedly, this is an observable effect, but it may also be underestimating the human proclivity for reproduction. Sex will doubtless survive, but as so many science fiction writers have surmised, the "thrill" of forbidden sex as we have known it may soon be gone indeed. But surely we are up to finding something far better to replace it with.

When all of the harmless forms of unusual sexual expression have achieved acceptance, there will still be other forms remaining that may truly be classifed as "perversion" and "morally wrong". These are the more violent Uranus sexual expressions that actually bring physical or mental harm to their unwilling victims.

Here the word "unwilling" is the key, as many forms of sado-masochism are equally pleasurable to the sadist and his very willing masochist. But such is not the case with the crimes of rape, sexual torture and enslavement, child molesting, and the like.

The elimination of the social stigma upon harmless "perversions" may, indeed, cause the brunt of the Uranus effect in sexuality to fall into these harmful catagories. If society is not careful, that indeed will be the case. As so many anti-pornography groups fear—where will it all end?

In a well regulated society, however, such should not be the case. We have, for instance, many socially acceptable outlets for our Mars-oriented hostilities and aggressions—peaceful competitions of every kind such as sports, games, and contests. These channel and release our negative Mars energies without causing harm to others. Similarly, our Uranus tendencies in sexuality may be channeled by acting out fantasy and role-playing in mutually willing and controlled situations.

The natural attraction for the sexual expression of Uranus and the thrill and excitement that it brings can, if openly dealt with, bring a positive and creative influence upon the sexuality of the coming age.

XI
Uranus and Crime

Crime, like perversion, is often a question of social definition. Essentially, crime is committing acts which the society, as reflected by its laws, considers to be damaging to the social order.

Thus, in all societies, killing a person is considered a crime when that act occurs within the society, as it certainly tends to tear down the framework of the social order if repeated. The same act, committed in the framework of war is considered heroic and admirable. Conversely, marriage within a society is considered normal, but in many cultures marrying a foreigner or a member of another race is still considered a crime.

Uranus, therefore, must necessarily and naturally be associated with crime since the characteristic of the criminal is that he is different from the average citizen in that he *breaks* (a Uranus word) the law. Being different in and of itself does not constitute crime (though sometimes it may seem like it), but the criminal, by definition, is different as it is the mainstream of society that makes up the laws he breaks.

Since the criminal goes against society and often

takes extreme risks as a result, the same high-tension qualities of Uranus associated with sexual perversions, real or imagined, are equally associated with crime. This really applies to those who commit premeditated crimes only, which account for most lawbreakers. Certainly there are those who break laws they are unaware of or who are simply emotionally unaware that the law they are breaking has importance or necessarily applies to them. This latter category, oddly enough, can apply to a variety of criminals from storeowners who open on Sunday to deranged killers who simply murder without thought or who kill professionally.

The majority of criminals do not choose crime as their first pursuit nor do they feel terribly comfortable about it. The pressure of society sees to that, though it is all too often that same pressure which causes the individual to turn to crime when no other more acceptable outlet can be found.

Such criminals may be characterized as "desperadoes" and may often be found to have Uranus prominent in their charts. Some internal psychological or external social pressure suddenly catapults them into often violent crime and, more than likely, a violent end. Examples are such Western outlaws as Billy the Kid (Sun opposition Uranus) and Jesse James (Uranus conjunct Venus, square Moon), or more modern desperadoes such as Bonnie Parker (Uranus square Jupiter) and Clyde Barrow (Uranus conjunct Mars).

Although many people grow up in difficult surroundings, under harsh or unfair personal and social pressures, not all turn to crime in reaction. Most simply internalize the financial and emotional scars and pass them along to their children. In a way, it is the revolutionary quality of Uranus that leads an individual to fight the society he feels has threatened or rejected him. Underneath the sometimes cool veneer of the hardened criminal is always the harsh Uranian hostility of the child who saw his needs for attention and affection all too clearly while society and circumstance ignored him or fed him on hostility and abuse.

But not all criminals suddenly break through their social inhibitions into a spate of unlicensed lawbreaking. As in every other field, crime has its geniuses who learn their trades carefully and rise to great heights of power, fueled by the pressure and tension Uranus provides. The father of American mobsters, Al Capone (Uranus conjunct Venus, opposition Pluto) who nearly singlehandedly founded American organized crime is a prime example. His legacy is hardly Uranian, however, as the mob was to evolve from a rebel, anti-prohibition crowd to a Plutonian network whose main intent is not law-breaking in a Uranian sense, but rather law-ignoring, becoming a law unto itself. The conservative, established, well-dressed Mafia businessman of today bears little resemblance to the Uranian independent mobster of Capone's time.

Another Uranian aspect of crime is simply clear and precise thinking. Like any other strategist, the criminal must see his possibilities clearly and strike quickly and effectively, making a clean getaway before the law can close in, decidedly Uranian skills.

Conversely, the detective must have like skills in order to anticipate his adversary and pounce on him when he's least expecting it. The more brilliant the criminal, the more imaginative and ground-breaking the detective must be in order to catch him in the high-tension game of cops and robbers.

The rarified intellectual quality of cerebral, yet stark realism, which Uranus portrays is best represented in the nativities of the great mystery writers. These people are able to conjure up the twisted patterns of both sides of the criminal fence. Certainly Earle Stanley Gardner with his web of suspects and surprise endings typifies it with Uranus square Mercury in his chart.

Probably the greatest of all mystery writers was

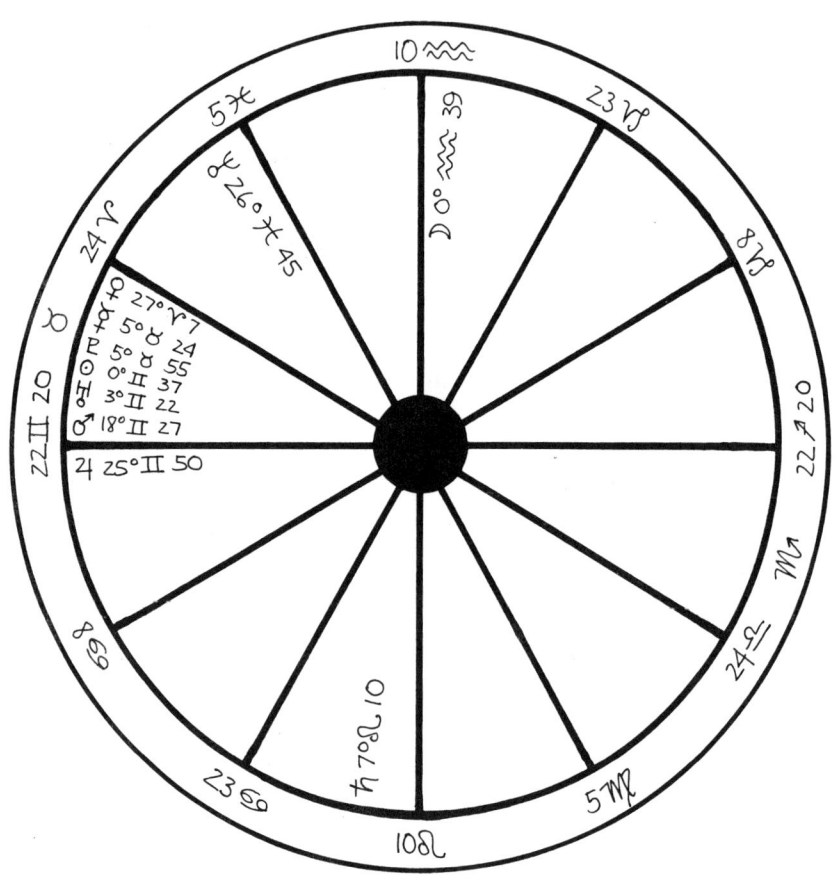

ARTHUR CONAN DOYLE

MAY 22, 1859

4:55 AM

EDINBURGH, SCOTLAND

the founder of the art, Sir Arthur Conan Doyle. His Sherlock Holmes and Professor Moriarty reached the peaks of incisive logic mixed with sudden flashes of inspiration and impeccable timing. Doyle, who had Uranus conjunct the Sun, ruling the MC and Moon (which rules the 3rd) found Uranian expression in more ways than mystery writing. He was one of the leaders of nineteenth century psychical research, believing that psychic phenomena were not intangible superstitions but concrete occurances that could be observed, catalogued, and studied in a scientific and clearly physical way.

Uranus is associated with crime, but not with all kinds equally. Uranus is there when the motive is desperate, rebellious, or the result of the response to internal and external pressures, resulting in high tension and its release in a form alien to the mainstream of society.

Neptune and Pluto both have their divisions of the criminal world as well. The forger, the embezzler, the con man—crimes in which the individual strays from the law by eluding or deceiving it rather than by aggressively breaking it—are Neptune's dark side. Pluto rules organized criminals, dishonest politicians, the high-placed briber, and those whose motive is not to break from society but to control it.

The Uranian roots of crime are the easiest to get at and do something about. Society can, with effort, be designed so that external pressures either from laws or upbringing do not force the individual to break from society. As always with Uranus these are clear and tangible causes and effects which can be changed and channelled. But it is a sure thing that the clear Uranian solutions to crime prevention will not be adopted until the more difficult-to-get-at Neptunian and Plutonian criminals who profit from and often run the perversions of our social system are swept aside.

XII
Uranus and Revolution

Within the framework of Astrology, Uranus is probably more closely associated with revolution than with any other concept.

The qualities of a revolution, political or otherwise, reflect many different sides of Uranus all blended neatly into one glorious uprising. Revolutions usually occur fairly suddenly, or at least are thought of in that way, although many important revolts devolve into wars of attrition that are seemingly endless. What is truly sudden about most revolutions, however, is not that they're over quickly, but that they are unexpected. If they were anticipated, they wouldn't happen to begin with. Either the necessary changes would be made more gradually, or the potential rebels stamped out ahead of time. Somehow, despite the warnings of clearer-headed individuals, the soon-to-be-overthrown order simply refuses to see, or is unable to see, the seeds of its destruction until it is too late. That is an essential ingredient in a good revolution. Had Marie Antoinette proposed a welfare program instead of saying "Let them eat cake", there would never have been a Bastille Day.

Aside from Uranian suddenness and surprise, a first-class revolution requires a manifesto, some statement that embodies the clear truth for which the rebels must fight and die; a program to establish and defend. "We hold these truths to be self-evident"—how Uranian can you get? Most people who have demands or grievances come to terms with them through give and take. A little compromise makes everybody less than happy, but less than fighting mad, also.

When there can be no compromise on issues (whatever they may be), then a revolution can occur. It is an all-or-nothing affair, more so even than most international wars. Rebels are seldom given the respectful treatment of a foreign prisoner-of-war; they are peremptorily hung. That is probably because there is more than property at stake in a revolution. Both sides are fighting for their principles which, unfortunately, are mutually exclusive. Men would far rather kill or die for an idea than for a possession.

Each side, particularly the rebels, believes it is fighting for a clear and untaintable truth that cannot be altered. It can only utterly prevail or be destroyed. Total Uranus.

A good revolution also needs leaders who can persuade their troops to view the matter in the way we have described. Most people don't see things in black-and-white, and would prefer an amenable compromise to a bloody battle any time. It is the duty of their leaders to stir them up and fill them with such indignation that they throw away caution, rise up, and wreck the social order.

A prominent Uranus would thus seem a necessity for such leaders and, indeed, it appears to be the case. Karl Marx had Uranus conjunct his MC, squaring Saturn. His mark on the world was the notoriety of his ideas against the establishment, not actual deeds of war. His later followers have a more personally active Uranus: Lenin (square Mars-Neptune conjunct disposing of the Moon and ruling the 3rd), Trotsky (in a T-square with

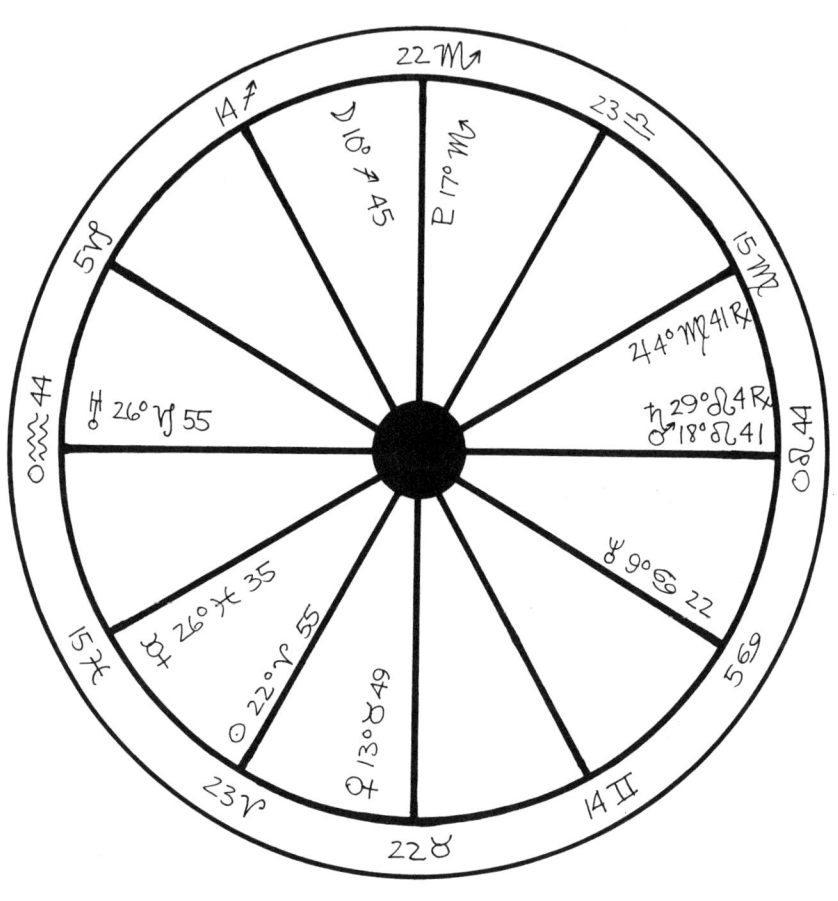

THOMAS JEFFERSON

APRIL 13, 1743

1:54 AM

SHADWELL, VIRGINIA

Jupiter and Mercury), and Stalin (conjunct Moon, opposition Jupiter in the 1st).

George Washington had Uranus square the Sun and ruling Mercury. Thomas Jefferson also had that same square, with Uranus on his ascendant as well. Certainly it is the leaders who must crystallize and articulate the urgent truth and necessity for revolt before anyone is going to get stirred up enough to do anything about it. The clearer the principles, the more courage and loyalty may be expected from the troops, even in the face of imminent defeat or death.

Such "clear truths" need, of course, to have a foundation in fact to be convincing. Without oppressing overlords, neither the U.S. nor the Russian revolution could have gotten off the ground. But such factual foundations need only be a take-off point. Equality for all was the avowed goal of the American Revolution, but at the time "all" did not include women, slaves, and, if Jefferson had had his way, non-landholders. Similarly, Communist Party members and officials wound up more "equal" than the rest of their supporters after the revolution was over. Such future details were overlooked at the time. If they had not been, both revolutions would have quickly ground to a halt.

That, however, is one of the problems of being individual and human, and the actual truth of any matter is different for every person. But you cannot build a revolution on that kind of disunity, so a generalized "truth" must be postulated and then sold to all individuals as a perfect fit, when in fact it is not. If there is reasonable enough truth therein, and there is enough discontent and hostility toward the establishment to get up a fighting spirit, that is all which is required. A good revolt can be effectively fomented.

We must come to terms with what we really mean when we attach the word "truth" to Uranus. One man's fact is another man's fiction. If everybody agreed on the truth, then there would not be revolutions to begin with.

Therefore, revolution may be said to be quite a lower manifestation of Uranian principles, embodying violence, precipitousness, and totality which occur exactly when opposing forces cannot agree upon the truth in any higher sense. The final opinion of which version of the truth is correct lies with the victor.

Mental or technological revolutions are less bloody, but they, too, reflect merely changing views of what the truth is all about. Newton's laws of the universe were "overthrown" by the more sophisticated views of Einstein yet before either were recognized apples fell to the ground and matter and energy kept in balance. Today far greater and more mind-illumining facts of the universe undoubtedly escape our notice, but those we do have serve us well as can be expected.

In the end, we must connect Uranus with what *appears* to be clear and unmitigated fact or necessity. The universe may indeed be all illusion, but it is certainly a very convincing one. The realities of everyday life, as we must work with them, are demanding enough to require our most careful attentions. Uranus characterizes these "self-evident" realities, and when others see them in a different enough light to come in conflict with us, we find ourselves becoming rebels.

Thus, many a desperate outlaw is as much of a true revolutionary as any country's founding fathers. The Uranian criminal works from as sincere and urgent a reality as the revolutionary; he is just alone and less successful. His acts are no more lethal or dangerous than those of an armed rebel; he simply does not receive the support of those around him. If he can gather enough support, like Robin Hood, from those who are not his victims, he can become a rebel hero. If enough people share his urgent needs and he is really skillful, he can become the father of his country.

Individual need, like truth, is relative, yet very real. When denied, it finds necessary violent Uranian expression. Depending on the situation, the time, and the place, it may then be labelled a heinous crime or a glorious revolution.

XIII
Uranus and War

In order to have a good political revolution, naturally you've got to have a war. There's nothing that can unify men in the cause of rebellion like the prospect of carnage, particularly if the slaughter is begun by the establishment in power. A few potential rebels hung for their political beliefs is sure to guarantee the deaths of thousands more by more violent and inspiring means—spears, clubs, swords, rifles, grenades, howitzers, rockets, bombs, and all the other Uranus-related paraphernalia of battle.

This is not to suggest that Uranus is a contentious planet. It is not. It is the combatants who are contentious, strictly a Mars-ruled proposition. Uranus is the ruler of their weaponry because its nature is precise and extreme, two fundamental principles of the planet. There is no arguing with a bullet, whether you agree with it or not: its reality is exceedingly clear, harsh, and sudden and bears no contradiction, only avoidance.

As rifle buffs are so fond of saying, "Guns don't kill, people do." That is true. Guns have no argument with anybody. But they serve as the total and irreversible solution to many an argument between people.

It takes Uranus in combination with Mars to produce any disagreement more lethal than a fist-fight —technology applied to anger, in other words. Mars provides the motive and Uranus provides the means. The combination of Mars and Uranus in conjunction is associated with violence and civil and military strife. In an individual horoscope it is considered to indicate one quick to anger or to rush into dangerous situations.

The Mars-Uranus conjunction is most interesting in mundane astrology, as it is said to mark the locations of upcoming world troubles where the two principles are involved. Riots, revolution, war, earthquakes, and violent disasters of other sort.

This is done by determining the time of the exact conjunction of the two planets which occurs every two years. The longitude at which this conjunction occurs at the midheaven and, less important, the ascendant, will be the focus of such disturbances for the following two-year period. The results of this are remarkably consistent and should command the attention of every astrologer.

For what it is worth, this conjunction can, in some instances, mark the end of major conflicts. For example, every war the United States has been involved in has ended within a month of this conjunction.

As might be expected, Uranus is involved with more than just the sheer violence of war. It is also involved with certain kinds of military tactics where sudden movement, surprise, or overwhelming force is required, particularly when involved with Jupiter.

An interesting modern example is the chart of the Arab attack on the Israelis in October of 1973, where Jupiter conjunct Moon was rising at the unlikely time of the assault (2 P.M.) with Uranus and Pluto beseiging the Sun (Israel) in the west.

Since quick, decisive action that catches the enemy by surprise is one of the main tactics in

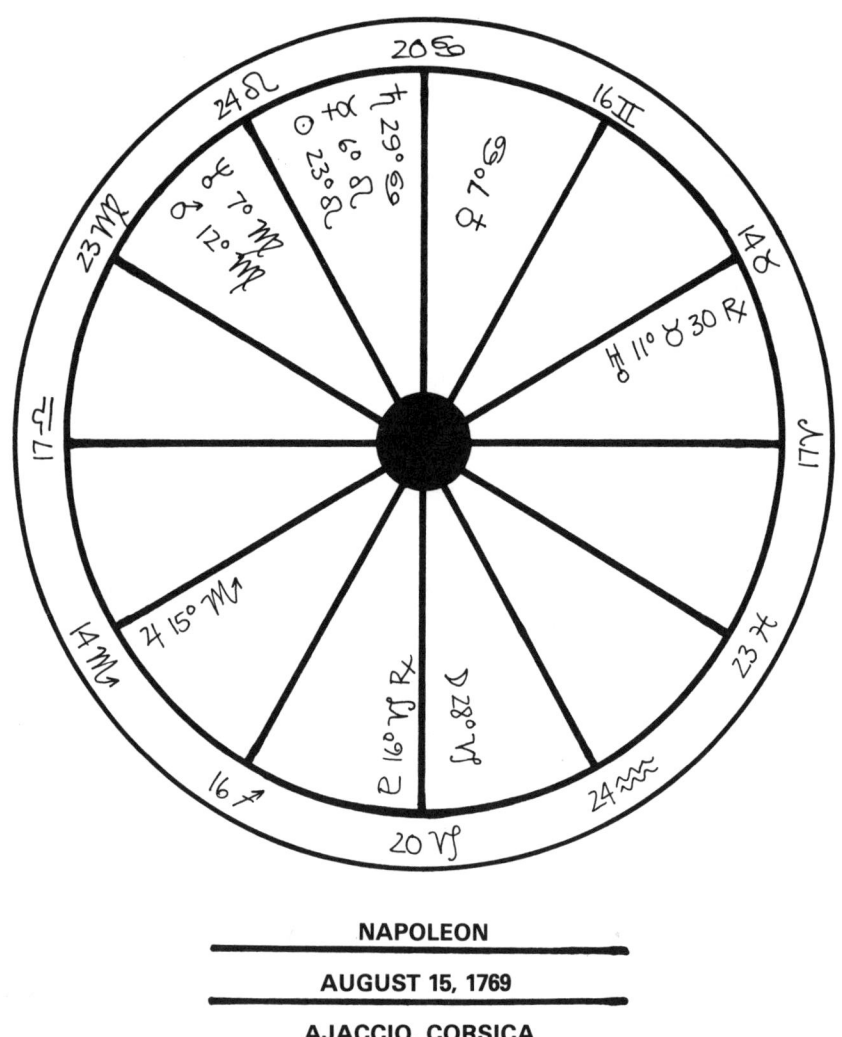

NAPOLEON

AUGUST 15, 1769

AJACCIO, CORSICA

traditional and modern warfare, one might expect a strong Uranus-Jupiter relation in the charts of generals who successfully use such tactics. Throughout history, that is the case, at least within the scope of reliable horoscopes (Julius Caesar's chart shows Uranus opposition Venus, but who knows if that is correct). Napolean had Uranus opposition Jupiter; Patton had Uranus in a T-square with Venus and Saturn; MacArthur, Uranus opposition Jupiter ruling the Sun; and Eisenhower, Uranus conjunct the Moon.

Of interest as well is Ulysses Grant's chart. His tactics were less to overwhelm the enemy by assault than by power of superior numbers and supply. This is reflected in his Uranus trine Jupiter. Ho Chi Minh, who managed to lead his tiny Vietnam most of the way to its victory over the most powerful nation in the world, chose quite different tactics. His chart, marked strongly by Neptune, suggests the strategy of constant evasion and shifting, elusive movement that so baffled the greatly superior forces accustomed to more forthright Uranian tactics.

Obviously, the most balanced and successful general in an all-around fashion should have elements of both Uranus and Neptune strong in his chart to be able to confound his enemies at every turn. Bonaparte is certainly the prime example. His chart shows Uranus opposition Jupiter and Mars conjunct Neptune, lending secrecy and subtlety to his movements as well as originality and surprise. As such he was the founder of modern warfare and indisputably ranks with Caesar and Alexander (for whom we have no reliable charts) as one of the three greatest generals in history.

Although it is not Uranus that motivates most wars, it is certainly the planet which envelops both the weaponry and the maneuvers that characterize what we generally envision as warfare. It also describes the very intellectual interest in the subject as a study by itself. Wargaming entails most of the Uranus aspects of war without including the Mars. Strategy, tactics, and the decisiveness of a quick battle or superior weaponry are all elements in the pursuits of an enthusiastic wargamer, whether an armchair enthusiast or an actual participant in practice military maneuvers. Only the anger and destruction of Mars is, fortunately, missing.

It is also that very Uranian fascination with war that leads us innocently into its horrors. Any generation who has not seen the pain and agony of war firsthand tends to easily develop an almost aesthetic appreciation of war's technical brilliance and decisive solution to the frustrating conundrums of the world. Indeed, war without its horrors is probably the most complex and challenging pursuits the intellect could grapple with. Certainly real war is the most total commitment of intellectual, physical, and financial energies of any pursuit known to mankind.

Man's fascination with organized armed conflict on the intellectual plane has led him into follies the price of which was misery and destruction beyond his worst nightmare. But war also seems to be a natural phenomenon, like earthquakes, hurricanes, and other Uranus-oriented occurrences. Some problems seem to brook no solution but to cut the Gordian knot. When diplomacy and understanding fail, the swift violence of Uranus sweeps all away, leaving a fresh field to start anew. Perhaps remedies may be found for these sad experiences that nearly every generation founders into, just as we may eventually learn to tame earthquakes and hurricanes. But it is far more likely that we will learn the control of our environment before we learn to balance out the conflicting goals within ourselves that cause us, time and time again, to let loose the dogs of war upon each other.

XIV
Uranus and Astrology

As we stumble into our so-called Aquarian Age, we hear more and more that astrology is the true Uranian science and that Uranus and Aquarius have rulership over the field.

One would have hardly thought that to be the case, looking at astrology at the beginning of this century. Many of the best astrologers were spiritualists and highly Neptunian in nature to say the least. Today, as then, much of a practicing astrologer's skill lies more in his ability to use his intuition, than in gleening anything definitive from a horoscope.

Astrology spans the natures of all the planets, just as philosophy does, as it attempts to describe every sort of person and event in a variety of both symbolic and specific ways.

One could hardly characterize astrology as being Uranian in the same, precise way as chemistry or physics. The amount of hard research comparable to those fields has been miniscule. Certainly astrology is based upon precise observations of planetary positions, but the behaviours these are supposed to encourage or influence have been put to the test far less often or as

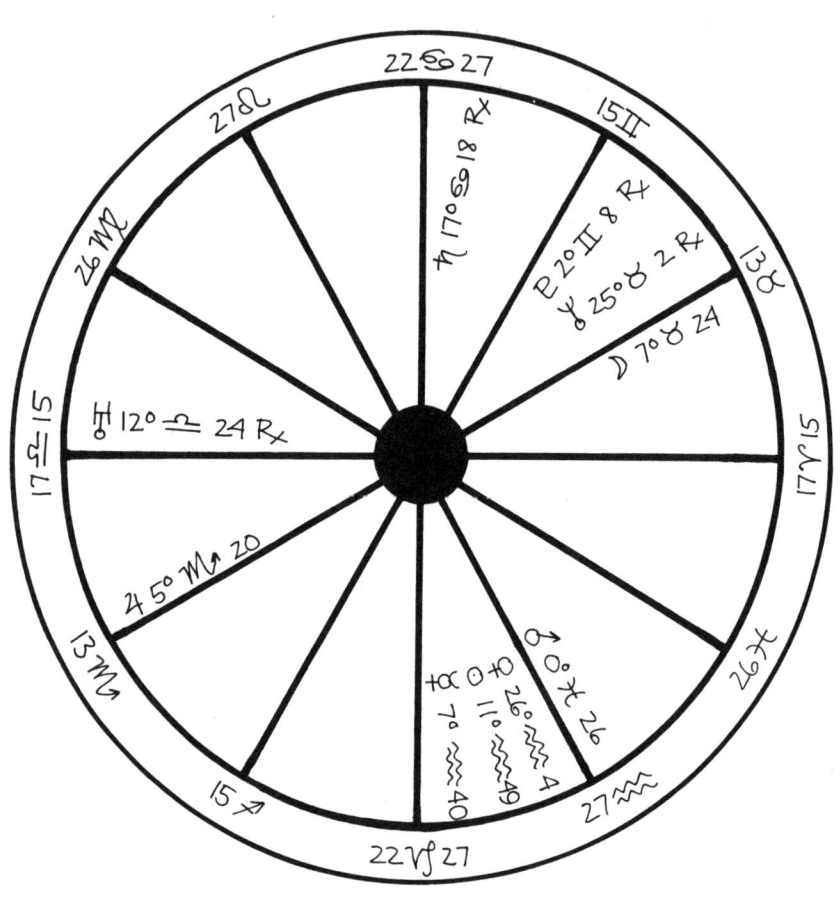

CHARLES E.O. CARTER

JANUARY 31, 1887

10:55 PM

PARKSTONE, DORSET, ENGLAND

precisely as supposed phenomena in other softer behavioural sciences such as psychology.

That heavenly bodies have a definite, measurable effect on behaviour, both human and otherwise, is certainly beyond doubt. What these effects *are*, exactly, is open to question and explanation.

Most astrological opinion has been based upon the individual experience of astrologers. This naturally has led to a widely disparate set of beliefs in the field. Astrological effects based upon joint observation of large numbers of cases have only begun to be studied. Those that are responsible for such studies are cautious in making any large or sweeping statements about the field, as scientists should be.

Yet hundreds of individual astrologers have not given up producing a seemingly endless spate of material on the subject, of which this book is an example, based solely upon personal experience and historical study.

This is not entirely reprehensible, however, as astrology is an interpretive art as well as a science. Its application depends upon the wisdom and understanding of the practitioner of the rich symbolism contained therein. As such, astrology is already a useful guide to the greater understanding of the human personality as reflected in the natal horoscope.

The prediction of events, long the desire and aim of many astrologers, is still in its infancy. Certainly general trends may be presupposed by looking at the transits to a given chart, but any consistently clear pre-description of events, even as accurate as an average weather forecast, is yet to be attained. That may come, however, as the weather bureau is far better funded and staffed by a very Uranian crew. There is still something to be desired in the accuracy of its forecasts, however.

As for the individual, a given practitioner may indeed be influenced in the style of his judgements and practice by a strong or a weak Uranus in his natal chart. An astrologer like Charles E.O. Carter with Uranus conjunct Ascendant ruling Sun, Mercury, and Venus, may tend toward the factual and research end of the field. Yet, similarly, someone like Dane Rudhyar may express his Uranus square Moon not in a technical way but in incisive insights into astrological character analysis.

Many astrologers like to paint themselves as "Uranian" or "Neptunian" in orientation, but this is all too often an underhanded apology for a lack of sufficiency in the opposite direction.

The talents needed for a good astrologer are both Neptunian and Uranian, because of the very nature of the field. Symbolism (the planets, signs, houses) by its very nature is general and pervasive, but the ability to crystalize the nature and effects of such symbols in any given circumstance is quite specific and precise. The understanding is Neptunian, the application Uranian.

Astrological research flounders without principles to direct it; and principles unfounded in reality are of no use whatsoever.

Therefore, the astrologer cannot be a one-sided person and be of any merit. Astrology attempts to describe, if not always explain, the events and purposes of the world in every conceivable area. Must not its practitioners strive to be that "Rennaisance" man who can bring breadth and understanding from one field to another as they share the astrological principles that relate to them?

Astrology, and the principles contained therein, represents a way of looking at the world that is at once balanced and dynamic. Above all, it is complete. It divides all of the world into a set of symbols described and numbered by the Sun, Moon, and planets. That there are currently ten bodies (8 planets, Sun, and Moon) to cover everything is merely transitory. The ancients got along well with three less (known) planets and other current astrological systems thrive using even more hypothetical planets than are known today. What is important, from the view of the practitioner, is the ability to transform symbol into substance and vice versa. That is the gift of the astrologer which will remain long after man has left the earth for other galaxies and planetary systems.

XV
Uranus in the Horoscope

Uranus' position by sign in a nativity has more than simply personal significance, unlike the inner planets and even Jupiter and Saturn. Since it takes seven years to pass through a given sign, it marks the style of a whole age group.

Thus, in the personal chart, it signifies, by house and aspect, how well and in what ways the individual relates to his peers or classmates. Teachers often comment on how a school seems to change its whole tone as one set of youngsters grow up through junior high and high school and another set comes along to supplant them. This is the effect of the change of sign of Uranus in the children's births.

As a motivating factor in the chart, Uranus represents the underlying will and purpose for discovery and enlightenment in life. In general a peer group shares an overall direction. Those significantly older or younger than ourselves seem to want something completely different out of life, as well they should with Uranus in a different sign than ours.

This difference is most noticeable in young people whom we see relating to the world very differently than thirty-year olds when the same age. In the case of older people it is much harder to judge, as we may indeed become more like them as we grow older ourselves. One must learn to separate what are the Uranus qualities of an age group from the natural universal qualities of the given time of life through which they are passing. By aspect, Uranus will tell how well the individual fits in with his age group particularly and with society at large in general. Those with easy aspects to Uranus or contact with benefics tend to fit in well with their peers and accept group opinions of what is interesting or important. Not that they are necessarily joiners or blind followers. Their interests just honestly tend to lie in the areas of socially accepted ideas and goals.

Those with hard aspects to Uranus or that planet in contact with malefics tend to be either social misfits, overachievers, or both. The tension produced by such aspects produces a personality that often clashes with the norm and is driven to explore new territory and taste forbidden fruits. This can have very rewarding effects, particularly later in life when the individual has learned to harness and channel those extraordinary energies, or it can lead to total disaster through general contentiousness, bad timing, and ill luck. Because of Uranus' basically extreme and harsh nature, it becomes most prominent and effective when in hard aspect, but its dangers are also greatest then as well. Individuals with Uranus so aspected must learn great self-control and discipline lest they go to extremes at the wrong time or in the wrong place. A strong Saturn is of great value to such people, while a weak Saturn allows Uranus to manifest itself in the worst ways as rashness, fickleness, violence, and impatience.

Properly handled, Uranus brings fire, conviction, and determination to the will. Like a fire, it needs to be banked lest it flare up and consume the individual. If not tended it will die out, leaving one cold and without light.

Whatever its aspect, Uranus represents the underlying will and goal-orientation of both the individual and his peer group. As has been said before, we all want the same thing, we just have different ways of going about getting it. Though our underlying currents may be going in the same direction, some reach fulfillment in the end while others get caught in the eddies of circumstance and drown. Whether one or the other happens is largely described by the natal position of Uranus and, equally as important if not more so, the transits of that planet.

Uranus in Aries

This position produces a generation of go-getters who rely less on brains or skill to succeed than on sheer force of personality. They are self-made people not in an economic sense so much as a psychological one. They have a singleness of purpose overcoming obstacles with overkill rather than simply going around them. Certainly they are people who have a noticeable and almost startling aura about them. They are often extremists, like Sarah Bernhardt or Vincent Van Gogh, or even criminals like Jesse James. On the positive side, this position lends great determination and strength of character as with Martin Luther King, Jr., or James Earl Jones.

Other examples: Friedrich Nietzche, Wyatt Earp, Shirley Temple, Jacqueline Kennedy Onassis, Ray Charles, Andy Warhol, Elizabeth Taylor, Ralph Nader, Oliver Cromwell.

Uranus in the 1st house

This position is traditionally associated with those who are odd or eccentric, though they would seldom label themselves as such. It does make for very original people with their own way of doing things, and who will do things in their own way, no matter what the pressure. It tends to make for a personality that is somewhat overwhelming to others, even though not intended to be. The individual will often be a loner both by choice and because his personality does not fit in comfortably with group situations. The manner may be somewhat brusque or short, and the tendency is to get business done with dispatch and get on to new things. This position often lends a high-strung temperament. If Uranus is very close to the ascendant, there can be an irregular frame and features. Such people are best off exploiting the unusual qualities of their personalities rather than trying to conform. Such conformity can never really be effectively achieved and it is better to be respected as an outsider than to be mistrusted as one of the boys.

Uranus transiting the 1st house
Particularly when transiting the ascendant, this can be a dangerous period. Accidents, falls, or any other kind of sudden mishap is more likely during this time. Of course you can't stay indoors for seven years until Uranus has cleared out of your first house, but it is wise to take extra precautions during the few months it actually crosses the ascendant. For the balance of the period, accidents are less likely. There will be a greater chance of one form of personality crisis or another. It is also a period of self-discovery marked by sudden and successive revelations which may not always be gentle or easy, but will be necessary for personal growth. Patience, which will be most needed at such times, is in short supply simply because that quality is the mark of an integrated personality and the period is one of *dis*integration followed by *re*integration. By the end of it you will have a very different and more cogent style of dealing with the world than before it began.

Uranus in Taurus
Just as the underlying style of Uranus in Aries is based on personality, the predilection of Uranus in Taurus is for expression through tangible principles and things. "Down-to-earth" is a good description for it. The tendency is to cut away the chaff and extraneous material and get right down to the basics. This may have

a very physical and political expression as with empire-builders Napoleon or Kaiser Wilhelm II, or just a down-home quality as with a variety of blues, rock, and "soul" singers (Little Richard, Aretha Franklin, Elvis Presley, Nancy Wilson, Tom Jones). This position has a disposition for loyalty and trust, as its values are centered around the physically real about which there can be no doubt. Even in the more ethereal realm of ideas one finds writers like George Bernard Shaw whose biting wit brought the airs of the British establishment down to earth. In music, too, is John Philip Sousa, whose marches are about as solid and earthy as any other composer in his genre. Throughout, the mission is to get down to reality, to bring supposition and speculation down to something reliable that can be depended upon. Sigmund Freud and Havelock Ellis did this for the then-unstudied field of psycho-sexuality. Uranus in Taurus represents a generation whose mission is to bring us back to basics and put our feet on the ground.

Other examples: Woodrow Wilson, William Howard Taft, Joseph Conrad, Theodore Roosevelt, Sophia Loren, Bill Cosby, Ringo Starr, John Lennon, Bob Dylan, Mohammed Ali, Barbara Streisand.

Uranus in the 2nd house

For those attempting to earn a steady living, this would be an unfortunate position, indeed. Therefore, anyone having Uranus here should attempt to derive his income from businesses which are erratic by nature. Areas like publishing or the music business where windfall profits are followed by no income at all are ideal for such a person. In any case, income is likely to be erratic as is the disposition of personal possessions. It is unwise to become attached to belongings, as they have a way of slipping away when one is most fond of them. Such a person should resign himself or, better, commit himself to a life of change and variety. If well adapted to this outlook he will fare well, for Uranus here not only divests one of possessions without warning, it also bestows new and unexpected goods to replace what is lost. This is a good position for a gambler or anyone engaged in the many professions available which rely on the gambler's touch. Once ensconced in such a profession, it is wise to rely on one's intuition as to which business deals to jump into and which to avoid, as this position brings considerable talent in picking the right choice based upon very little concrete information.

Uranus transiting the 2nd house

This brings a time of shaky finances and, of course, rarely happens to those with Uranus here natally, who would be best prepared to withstand it. It is a time to expect the unexpected and when one seems rich, his funds could be gone overnight from an unexpected and unavoidable drain. It is, therefore, wise to be very careful and hoard your money during this period, for you might suddenly have to bring your entire financial resources to bear at a moment's notice. The net result of all of this is to form a broader and wiser philosophy of handling your affairs, unfortunately through the teachings of experience. For those that highly treasure and cling to their possessions, this will be a period of trials, tribulation, and loss. For those who put little stock in goods, it will be an adventure and a variegated confirmation in their distrust of material things and their faith in other values. Insecurity is the keyword here, so do not make plans that you cannot afford to have upset. Forewarned is forearmed, at least to an extent, and this transit should enable you to laugh at the vicissitudes of life—after they are over.

Uranus in Gemini

As the essential drive of the person with Uranus in Taurus is to the earthy and substantial truths of life, those with Uranus in Gemini seek truth in the insubstantial realm of ideas. Verbal expression of a thing is essential to grasp its reality. St. John's phrase "In the beginning was the Word" is the ultimate western expression of this way of looking at things. Matter is only made real by the grasping of its inherent conceptual form and description. Such people are not too happy in the everyday world. The common affairs of men seldom conform to more lofty ideals of truth and justice. But the person following these too closely is usually doomed to frustration and failure. Only those who can use their rhetorical idealism to inspire others, while keeping their own feet on the ground concerning the realities around them, can use this position of Gemini wisely and effectively. Otherwise it inclines to burn itself out and place the native in the hands of the very injustice he seeks to rise above. It is thus a better position for the scholar, the researcher, or the writer rather than for those in the world of business and politics. Examples: Charles Lamb, Jane Austen, Daniel Webster, Pierre Curie, Arthur Conan Doyle, Billy the Kid, Anton Chekhov, Gustav Mahler, Alan Leo, Alfred North Whitehead, Rudolph Steiner, William Randolph Hearst, Henry Ford, Paul McCartney, George Harrison.

Uranus in the 3rd house

It is often said that this position of Uranus leads to eccentricities of the mind. Whatever that may mean, certainly this position inclines the native to thought processes which are different and more instantaneous than those of others. Where another person might follow all the logical steps of a problem to its end, this position will lead the native to leap directly to the conclusion or get to it by some unexpected short-cut. Unlike the more pedantic thinker, however, a problem will either offer him an immediate solution or simply appear altogether unapproachable, depending upon whether inspiration has struck or not. Thus, such natives should avoid areas of endeavor where much verbal repetition or regularity is needed as, for example, in schoolteaching. It would be better to follow a profession where a steady flow of ideas is less important, compared to an occasional brilliant one, such as advertising. The personality tends to be a bit restless and that same propensity to suddenly hit the nail on the head, when ill-aspected, leads the native to constantly suspect that the real answer is lurking about somewhere and what he has currently got is not of much use. This is a good spot for the "idea man" but not for the person who develops ideas and turns them into concrete realities.

Uranus transiting the 3rd house

This is usually an excellent period for learning, and those who go through it in an earlier period of life when their minds are more open are the most fortunate and usually benefit the most. It denotes the continual opening of new worlds of ideas and knowledge that are different and contrary to what one has known before. For those more set in their mental ways, this can be a difficult period as new facts and events spill into their experience. This can shake up and contradict the intellectual certainties they thought they possessed. It is best to be quite open and passive of mind during these years so as to absorb and benefit from new experience rather than be buffeted about by it. On a more mundane level, this is a period when the general organization of everyday life goes awry, such as meetings, plans, phone calls. This is just a reflection of the overall turbulence Uranus brings into the realm of thought at this time.

Uranus in Cancer

The outgoing, revolutionary quality of Uranus does not seem at first to blend well with the indrawn, conservative nature of Cancer. But this is merely a matter of direction. Uranus represents the area in which the native seeks his ultimate truths, and is willing to make his ultimate sacrifices. Those with Uranus in Cancer will defend the home fires to the death, because that is where they see the most important values to lie.

Here you will find patriotism not of an agressive, outward kind but of the sort that is motherly and protective of the national hearth, represented by such as Queen Elizabeth I. Where the effect is outgoing, it is expressed in finding new lands to become "home", or homesteading, such as with Davey Crockett. The native will not necessarily stick to one dwelling, as the restless quality of Uranus impels him to ever seek the ultimate and final Home. But whatever the current abode is he will fortify as no other. If his chosen home is not in his own possession or control, such as with Nikolai Lenin, he will seek to do this. Lenin's politics were Marxist, but his goal and accomplishment was Mother Russia and the ultimate restoral of the land to the people. Such ennobling of the home country was expressed eloquently in Rudyard Kipling's paeans to Britian, and profoundly, in Marcel Proust's belief that the man who remains at home gleans as much experience from the world as the most adventurous traveller. Again, the home-building need may be quite physical as seen in the work of Frank Lloyd Wright. These are all outstanding examples, but the ill-aspected Uranus here may just as well devolve into contentious bickering with a mate over the condition of the living room. Uranus is seldom easy in Cancer because of its dissimilitude with the sign and therefore may be expected to manifest itself more explosively than in others, for good or ill.

Uranus in the 4th house

Those with Uranus here usually make a rather marked change in lifestyle from that with which they were brought up. They are generally dissatisfied with their parents or community and this expresses itself throughout their lives in a tendency to periodically radically rearrange their surroundings. The life of a traveller or one in a profession that necessitates moving about a great deal is well suited to such a person. But, just as they tear up their roots with a will upon moving, so, too, do they eagerly ensconce themselves in their new domain. In the realm of the intellectual nature, this is a good position for those in the business of home improvement design, though not in actual construction, which requires more staying power. Its most positive side is the ability to root out and overturn false premises and ill-laid foundations, both physical and intellectual. The adverse aspects are seen in those who attack established ideas out of sheer contentiousness.

Uranus transiting the 4th house

This is almost always a period of travel and wandering. It is most difficult to stay in one place. Even if one manages to stay in the same city, it seems as if one must move from one dwelling to another due to unforeseen circumstances. It is also often a period of emotional trial, particularly for children who are most upset by the lack of a stable home. For the adult it may be marital breakup that causes emotional stress, or career changes, or even those vagaries of fate such as the burning down of a house or the sudden acquisition of a new and better one. It is well to resign oneself to the adage "home is where you hang your hat", as it is never so true as during this period. This transit can have the benefic effect of teaching one that home is truly where the heart is, not where the house is. If this is learned, then the transit may be said to have been a success.

Uranus in Leo

Because of its intensity and fire, Uranus is well suited to Leo and can find an easy outlet there. This generation seeks its fulfillment through the expression of warmth and generosity, wherever such qualities may be applied, whether on the personal or the social and political level. In politics we find such grand humanitarians as Winston Churchill and Herbert Hoover. In the art of poetry this may produce a Robert Frost or Carl Sandburg. Throughout, the feeling is broad and the tone of life that of expressing scope and grandeur. Uranus here may produce the musical breadth of Pablo Casals or Maurice Ravel. It may also be a personal commitment to helping mankind, as seen in Albert Schweitzer. Uranus is at its best here when able to paint life on a large grand canvas, protraying *all* areas of life. Harsh aspects to Uranus here may find well intended royal ambitions thwarted, as with Mary Queen of Scots, or master craftsmanship brought to an untimely end, as with Houdini. But left to finish its work, Uranus will cover all of the ground. Fine examples of this are father of film directors, D.W. Griffith, and free spirits of the stage such as Lionel Barrymore and Isadora Duncan. Massive and definitive achievements were made in the spiritual world, Neptune's domain, as shown by the monumental life works of Edgar Cayce. The generation of Uranus in Leo creates with a bold stroke and is turned awry only when limited by a world that lacks its vision.

Uranus in the 5th house

This position inclines the native toward extreme physical desires. These may not always be actively expressed. The pursuit of pleasures, whether bodily or artistic, will be a main motivating factor. However, given the quality of the planet Uranus, such enjoyments will tend to be sporadic and therefore must not be held on to. This is an ideal position for the gambler who can quit while he's ahead, and the worst for one who cannot. Social company will change, and should be enjoyed but not relied upon. Here Uranus is in a good position for entertainers and those who rely on brief but intense public performances for their living. It can, when in hard aspect, lead to intense sexual abnormality on the one hand or to the spiritual deification of sexuality on the other. Uranus here tends to make one keep a smiling face to the public, as the inner drive and belief is that happiness and enjoyment is the proper state of things even in the worst of circumstances.

Uranus transiting the 5th house

If one were to spend a night in a giant amusement park, it might seem that there are just too many opportunities for thrills, rides, and fun to possibly take advantage of. The next day, in the humdrum world, there is nothing at all. These two extremes are a description of the years during which Uranus transits this house. Either there is too much enjoyment for the senses to handle or there is a sensual desert. Tastes for the bizarre and unusual are awakened and the individual should be prepared to recognize in himself those needs and desires he previously relegated to the peculiar or abnormal. For the hide-bound, this may be a disturbing time, but for most it can be a time of liberation from the chains of an upbringing overconstrained in matters of sexuality and in just the everyday art of having fun.

Uranus in Virgo

This sign seems to have little in common with the extreme and explosive nature of Uranus. However, in our century it has found expression in many accomplished artists, politicians, and other public figures. Uranus in Virgo finds its truth in the delightful details of life, those homey little expressions that turn otherwise strangers into just folks. Comedian Will Rogers was the master of just such conversion. By being specific, one touches upon reality in a way that more intellectual or more mental concepts cannot. The height of over-specific comedy is reached in the complex precision antics of Mack Sennet slapstick or the elaborate foolery of Rube Goldberg's cartoon constructions. Uranus in Virgo is concerned more with technique than with substance or interpretation. This is well illustrated by the over produced extravaganzas of Cecil B. DeMille. No one in the arts ever made so much of the introduction of one technique as Pablo Picasso. But, as Virgo is also concerned with physical nurturing, we find people such as Franklin and Eleanor Roosevelt who did more to actually get down to the details of helping humanity as a whole than anyone else in their time. So, too, we find fascination with form and detail in the literature of Vachel Lindsay, Kafka, Joyce, and Virginia Woolf. Even in more ethereal realms there are the endless volumes of esoteric and spiritual teachings of Alice Bailey and the crackling precision of L.Edward Johndro's technical astrology. In less advanced nativities these qualities can be expressed as most nit-picking obsessiveness and short-sightedness. At their best they glory in the fine particles which, in their perfect imperfection, are the building blocks of society and of the Universe.

Uranus in the 6th house

This is not generally considered a good position for health. It causes rather sudden onsets of undefinable ailments, such as nervous disorders, or unexpected accidents. Such people really do not suffer any more than others, but the nature of the afflictions and the anxiety experienced anticipating them adds as much grief as do the problems themselves. It is therefore best to simply proceed blissfully through life and not worry until a problem presents itself. One cannot prevent the unexpected, and one should not try to. Such attempts are an exercise in frustration and futility. This is also a good position for the native to take care of the details of his own affairs as much as possible because secretaries, assistants, and the like will prove to be rather unreliable.

Uranus transiting the 6th house

For those who rely for employment upon others, jobs will come and go in the most unpredictible fashion. But, as soon as one opportunity has dried up, another will mysteriously present itself. Similarly, employers will find their employees disturbingly unreliable during this period. But, again, help will come from the most unlikely quarters to set things straight again. It is, in general, a challenging time that teaches the native adaptability out of necessity and leaves him with a much broader base of command and potential. Certainly health problems may be expected to arise, but there is no point in worrying about it beyond a good checkup and a reasonable insurance policy. In the end, the lesson of this transit is one of learning one's dependency upon one's fellow man and the increased mutual respect and love that must bring.

Uranus in Libra

To say that those with Uranus in Libra are overconcerned about justice and fair play might seem like a cliche, yet it is true. One might also say they are inclined toward reshaping and rebalancing things to suit their taste. Here we find such moralists as John Stuart Mill, Nathaniel Hawthorne, and Sinclair Lewis, whose main direction was their own personal brand of right, though it did not always coincide with contemporary society's view. We see a similar even-handedness in the arts in Irving Berlin, Marc Chagall, and Maurice Chevalier. This drive to put things into a personal view of proper proportions can be good or bad depending upon the chart. Witness Hitler's maniacal drive to reshift the balance of the world compared to the tender and gentle Charlie Chaplin born only four days earlier. Whether in war (Patton, T.E. Lawrence) politics (Disraeli, Eisenhower, DeGaulle), storytelling (Hans Christian Anderson), or even astrology (Marc Edmund Jones) Uranus in Libra has a kind of evening quality about it quite often mixed with charm unless most evilly aspected. Then it tends to go about upsetting things. But good or bad, its direction is to put things into proper balance and proportion as the individual sees it. This will be the main quality marking his style and particularly the style of his generation. But whether the goal is, in the eyes of others, actually bettering things or simply stirring them up, the means used to gain that end will most often be charm and persuasiveness, since this position considers those qualities ends in themselves as well as potential stepping stones to achieve other effects.

Uranus in the 7th house

As far as personal relationships are concerned, this is usually a troublesome position, unless the individual is truly a loner and can take and leave his partners as he finds them. Uranus here brings sudden impromptu marriages and equally as sudden divorces. Even for those with seemingly stable partnerships the union may be unexpectedly interrupted by the force of circumstance pulling them apart, either through the death of one or the financial or political necessity of living apart. Certainly the native will not only marry but have business dealings with those quite unlike himself so that no one could ever call them "two peas in a pod." Thus we find interracial marriages, wedding of rich to poor, short to tall, intelligent to dim-witted and the like. With this position it is best to simply expect and enjoy the unusual in personal relationships and not require too much normalcy or stability of them. The greater the individual's self-sufficiency, the happier his personal life will turn out.

Uranus transiting the 7th house

This is a rough time in most people's lives, particularly those who must depend upon close working relationships to support them either emotionally or financially. Flare ups between husband and wife may be expected and they are damaging to the extent that each is not a whole and self-reliant individual. For unattached persons it can be a very exciting period of short lived, but intense romances and bittersweet love affairs. The lesson here is loving one's partner freely without leaning too heavily upon the relationship for emotional support. The same goes for friends and business relations where one finds that good fences do, indeed, make good neighbors.

Uranus in Scorpio

If anything characterizes Uranus in this sign it is an intense singleness of purpose and the desire to get to the very depths of a problem or subject. People with this position tend to run a bit on the "heavy" side: Wagner, Immanuel Kant, Cesare Borgia, and Immanuel Velikovsky, for instance. They tend to be fairly self contained and go about their purpose with a will, such as the news gathering of Lowell Thomas or the money gathering of J. Paul Getty. This self-retentiveness can be found in fields as disparate as muscle building (Charles Atlas, who developed isometrics, entirely inwardly-oriented exercises) to nation building (Mao Tse-Tung, who developed China, an entirely inwardly-oriented nation). Not everyone born during the seven years Uranus is in Scorpio is necessarily a monomaniac, of course, but the drift of the generation is to take matters fairly seriously and plumb them to some considerable depth. Even in comedy the slant is either very hard hitting (Groucho Marx) or somewhat sexually obsessed (Mae West, James Thurber). The greatest asset of this position is its self reliance and completeness. There is an ability to use all available means to their utmost extent as Segovia did with the guitar or Buckminster Fuller did with environmental design. As a result, such persons often tend to be loners, as their main direction is inward and bent on achieving their own fixed goals rather than relying on interaction with others for a sense of purpose. In their case, the steadfast journey toward achievement is its own reward. Other examples: Amelia Earhart, William Faulkner, Aldous Huxley, J.Edgar Hoover, Henry Miller, Isaac Newton, Copernicus.

Uranus in the 8th house

This position lends a liking of antiquities, history, ancient languages, and the like. It may equally lead the native toward other areas just as obscure or difficult to really get at, including the occult. Uranus here serves as a spark of light in the dark corner that is the eighth house and may shed new light on many hitherto unknown subjects. Thus, such individuals are willing to leap into new areas of exploration or endeavor knowing full well that they will be the first to go there and have little idea of what awaits them. It is the position of the explorer, not in the physical but in the psychic sense. As such, the psychic powers are heightened here and accurate premonitions and other phenomena are not uncommon.

Uranus transiting the 8th house

This is a period when all sorts of unexpected and often disturbing insights come unasked for out of the blue. It is as if the dark corners of our minds suddenly pop out at us in full and surprising clarity. It is good for those wishing to hone their psychic abilities or become more acquainted with occult matters, though not through formal study but by revelation or insight. On a more mundane physical level this can be a period of interrupted or irregular income and may also be characterized by occasional unexplainable sexual dysfunction. Unlike most other transits of Uranus, its effects are of a rather non-specific, non material kind that are personally meaningful yet hard to pin down.

Uranus in Sagittarius

This is a very comfortable location for such a harsh planet and it can produce some pretty relaxed people, such as Bob Hope, Bing Crosby, and Arthur Godfrey. The source of that quality is an overall concern for mankind that marks Uranus in Sagittarius. The life expression of this generation runs towards congenial philosophizing on an everyday level and towards universal humanism on a higher level, as may be found in the works of Whitman, Melville, Steinbeck, Thoreau, and Hemingway, all of whom shared this position. In music and the arts it lends itself toward eclecticism, and in the music of Gershwin or Copeland and the films of

Hitchcock or Disney. This same universalism is shared by simple down-home philosophers like longshoreman Eric Hoffer or trumpeter Louis Armstrong and by high political rhetoriticians such as Karl Marx and Friedrich Engels. All, in their own way, share a common human goal, that of humanity itself. Even those whose Uranus qualities go to thwart society at large do it in a grand style, such as Al Capone. Uranus, which can be so dictatorial in other signs, here becomes democratic. It does not express the arbitrary egalitarianism of Aquarius, but rather a warmth and understanding which recognizes each individual as a whole and unique expression of mankind. This enables Uranus to freely express its fire, and takes the jagged edges of intolerance and extremism off the usually harsh planet to replace them with a wider vision of the various truths that describe mankind.

Uranus in the 9th house

This is an excellent position for Uranus, bestowing upon the native a deep sense of religious faith, not necessarily the church-going kind, but that sense of universal unity on which true spirituality is founded. The native is frequently eccentric in his manners and opinions, as he bases these upon his own perception of reality and not upon arbitrarily accepted norms. Naturally, this position is associated with distant travel, particularly voyages of discovery, as with Admiral Byrd who had Uranus here. Those with Uranus here tend to be quite intuitive and sensitive to the dictates of their subconscious as well as to the emotional state of those around them.

Uranus transiting the 9th house

This is often a time of wandering or travel, somewhat erratic in nature, and is always a time of extensive learning from outside circumstances. Many of the native's long-held views of the world will be shaken up and rearranged and his outlook on humanity will be considerably broadened. This is, except for its irregular nature, perhaps the best transit of Uranus of all the houses, allowing that planet to do what it does best: deliver knowledge and truth. This learning is irregular in the sense that knowledge derived during this period is unstructured and must be accepted, where and when it appears. For the alert observer, it is a time of rich, varied experience and education provided free by the world around him.

Uranus in Capricorn

Uranus in Capricorn is a sure signifacator of the dogged, do-or-die fighter. Its inclination is to take up the most difficult cause and carry it to success against the most impossible odds. Success is not usually gained through brilliance of strategy but by an untiring persistence that eventually wears its opponents down. The best example of this is the rather unoriginal but unflagging Grant's triumph over the military genius of Robert E. Lee. In ancient Rome we have Cicero, who beat down his opponents by that endless oratory which students of Latin so dread having to wade through. The cause may vary, but the style remains: Clara Barton, whose life-long struggles founded the Red Cross, Walter Reuther, tireless friend of labor and bane of management, Dag Hammarskjold, whose unending quest for world peace led to his death in an African jungle, Louis Pasteur, whose patient and persistent research revolutionized medicine. When they are not aiding a great cause, however, natives with this position tend toward the conservative, holding the philosophy of the self-made businessman who has the right to keep all he can acquire: Howard Hughes, Aristotle Onassis, Ronald Reagan, Barry Goldwater, Ayn Rand. This is not a particularly original sign for the arts, except where there is need for the arts to battle accepted social norms, as with Baudelaire. Usually it will go to produce art or music that is an "old favorite" in its own day, such as the melodies of Stephen Foster or Johann Strauss. True originality seems mainly to be found when the artist must struggle against some social or physical handicap, such as Django Rheinhardt, to get his work across. However it may be expressed, the message of this generation is a belief in hard work and the eventual victory of sheer spunk and determination over the greatest of odds. Other examples: Jean-Paul Sartre, Simone DeBeauvoir, Lyndon Johnson, Tolstoy, Mary Baker Eddy, Errol Flynn, John Wayne, Charles Jayne.

Uranus in the 10th house
Those with this position will find it hard to keep out of the public eye for one reason or another. It denotes an unusual reputation, and all the problems that brings with it. Rises and falls in the career will be abrupt and extreme, but not permanent. The most eccentric and unusual side of the personality is· on display to the public. This aspect may be considered the root of the native's success. The positive and negative response to such a personality tend to be extreme. Such a person attracts fierce loyalty and rabid hatred, both rather blind in nature. Because of the stormy nature of this position it is well to choose a career which does not require long and steady growth, but something more fickle and risky like politics or some forms of show business where the wild swings of Uranus can have full play and be taken advantage of rather than being a liability. Whatever the difficulty, the native should never despair, as he is the child of fortune and while foundering in the storm he is but a few moments away from being cast ashore onto some fertile island of fortune, from whence he shall begin a new cycle.

Uranus transiting the 10th house
This is likely to be a financially, and sometimes personally stormy period. This is true particularly for those whose emotional stability is linked with career stability. The nature or location of the career may

change several times in a short period, and those who will most benefit from this are individuals with a high degree of adaptability. Patience is required to weather these years as there are likely to be various untruths circulated concerning the native's reputation. This can only be counteracted by as much steadiness of character as can be mustered and demonstrated. Because of the sudden shifts and turns of employment, however, the native will discover he has many previously undiscovered and useful talents that emerge from hiding in the midst of travail. The vicissitudes of this period teach both a distrust in determinism and a greater faith in self and the powers that protect us in time of need.

Uranus in Aquarius

Uranus is said to rule Aquarius and certainly this planet is at home here, but often so comfortable that its explosive greatness is dissipated. Throughout, it offers appeal and communication to the common man and those with this position often meet success in exploiting the whims and needs of the general public. The wide, grass roots appeal of John F. Kennedy represents this mass attractiveness at its best, as does the gentle if organized appeal (thanks largely to Watergate) of Eugene McCarthy. At its worst this placement can reach the innermost hatreds and prejudices of the public and stir them up, as seen in former vice-president Spiro Agnew. On any level this position successfully speaks to the man-on-the-street. Thus one would expect it to turn up in people involved in the news media, as it does with TV network anchormen Walter Cronkite and Howard K. Smith. And who does the housewife turn to when she wants some good common sense advice? Ann Landers and Abigail Van Buren ("Dear Abby"), of course. Some good grass roots, middle-of-the-road Christianity? Billy Graham. Classical music for the millions? Leonard Bernstein. What is alarming is how dull and commonplace these characters seem to be, but then so is the stereotype of man-on-the-street or the average housewife doing her chores to the accompaniment of the TV blaring the midday soap operas in the living room. The genius of Uranus here is in the discovery of the special value of the average person, upon whom all society and government is based. This talent is of particular value to society when possessed by men of uncommon means or position, such as Andrew Carnegie, who gave away so much money for the cause of bettering the lot of the common man. Even today, Henry Ford, Jr. remains one of the few high corporate executives to effectively have his eye on more than just the good of his own company. In arts and literature, this generation has its attention on direct appeal to people rather than to some higher intellectual cause. Mark Twain, Tennessee Williams, Arthur Miller, and cartoonist Walt Kelly ("Pogo") all speak to and about the average person and his joys and problems. Though lacking in the violence, struggle, and triumph usually characterizing Uranus, this position perhaps represents the truest view of what really holds the world together.

Uranus in the 11th house

This position relegates the native to a particularly unusual and intermittent kind of friendship. He will tend to make fast friends very suddenly and equally as suddenly drop them. Or he may find his closest friends very peripatetic and only see them occasionally. At any rate, he will find close personal friendships to be very intense. But he must take care against betrayal as he is gullible, may place great trust in a secret enemy without realizing it, and then feel terribly bitter about it afterward. This is a good placement for making connections with those in higher positions who are able to give financial help or aid. Indeed, they will appear to drop in out of the blue when least expected. The best approach to Uranus in this house is for the native to try to stand back a bit from close personal relations as he tends to be somewhat unreasonable, both positively and negatively, concerning them.

Uranus transiting the 11th house
This is a topsy-turvy time concerning the positions of those around you. The staunchest friends may desert you in time of need, while strangers and even those seeming to be enemies rush to your side to aid you. It is a personal sorting-out period, and it is wise to try to depend upon no one but yourself if you wish to be sure of getting consistent results. It is an exciting period, however, as it tends to significantly broaden your range of acquaintances, putting you close to people you might never dreamed of associating with before. As long as reasonable self-reliance remains, it will be a period of great deepening of personal and human understanding, both of yourself and particularly of others.

Uranus in Pisces
This is a most curious and uncomfortable position for Uranus, as the qualities of the sign are quite opposite that of the planet. This results in a great deal of difficulty in putting one's finger on a life purpose and the internal pressure that brings. When finally expressed it tends to be quite violent and explosive, as if the heat of Uranus has boiled the cool waters of Pisces and produced a steam explosion. When manifested in the occult and mystical side of Pisces, it produces visionaries who are far ahead of their time and quite out of step with society. Such men were the astrologer-psychic Nostradamus and artist-visionary William Blake. Here the psychic forces came rushing out of them like a flood, to the consternation and disbelief of those around them. Despite the truth and wisdom of their visions they were of necessity social outcasts, being so far separated from the vision of their

time. Similar is the case of Mozart, whose remarkable music poured out of him from the age of five and yet, despite its greatness and the acceptance given him in his own time, he met an impoverished and sorry end while only in his thirties. General Lafayette, who did so much to help a fledgling America in its war of independence met a similarly unjust fate at the hands of his own countrymen. On both sides of the French Revolution we find the worst sides of Uranus in Pisces—in the harsh weakness of Louis XVI and his wife Marie Antoinette, and in the capricious violence of Robespierre who finally died by the same guillotine as the unfortunate royal couple whose deaths he helped arrange. The contradictory energies of Uranus and Pisces may be creatively channeled, however, as witnessed in the watery paintings of expressionists Monet, Cezanne, and Renoir and sculptor Rodin, or in the vascillating music of Tchaikovsky which is alternately quite watery and sentimental or violent and glorious. This placement is seen in the somewhat moody works of William James or writer Thomas Hardy, in which the hard facts of reality are seen to be resting on the shifting grounds of ineluctable fate. The personality here is bound to be dual, and the native must learn to express both sides of it freely. It is mainly in the suppression of one side or the other for any length of time that the danger lies.

Uranus in the 12th house

This is a dangerous yet exciting position for Uranus. It signifies sudden illuminations from the subconscious and the non-material world, but it also bodes ill from secret enemies who may attack the native with impunity. It lends a certain quality of suppressed violence and hostility as well, because the native seldom finds the proper opportunity to let off steam. It is beneficial in that it aids in solving the most difficult and obscure of problems and sheds light on areas often unknown even to exist by others. It may also lead the native unwarily down roads where none may follow and where he may be cut off. It may figure equally in the chart of an assassin or his victim. Well-aspected, the native may become unsuccessful in areas of secret operations or remote and obscure projects. Ill-aspected he may land in prison or some other institution of confinement or fall victim to an enemy he had trusted as a bosom friend. On the spiritual level, it has all the implications both of danger and discovery, of exploration into the highest (or lowest) areas of existence. The magnetism of the unknown may show the native heights of transcendant ecstasy or suck him down in a vortex of destruction. Great care is indicated, but great rewards are to be had.

Uranus transiting the 12th house

This is a rather murky period, regarding the effects of Uranus, and more incipient than immediate in its effects. Certainly, strange dreams may be expected and sudden emotional upwellings that never quite reach the surface. Great attention should be given them, however, as they presage the personality upheavals that will follow when Uranus passes over the ascendant into the first house. It is an excellent time for exploration of occult matters, as long as it is done with care. It will have both the effect of shedding light on hard-to-reach twelfth-house matters and at the same time making it easier for unknown dangers to manifest themselves. But most of all, it is a crucial time for self-observation, as many of the personality patterns most stable and accepted will at this time be the ones to crack and fall. It is important to remember that the groundwork for the abandonment of ingrained personal habits and assumptions should be laid at this time.

Transits to the planets and lights

The Sun

The transiting conjunction of Uranus to the natal Sun is likely to mark an important time of personal and psychological upheaval. This period can be either very creative or very destructive, depending upon the resiliency of one's character and the surrounding physical circumstances. This transit causes a breakthrough in the internal barriers and illusions. This, ultimately, is a positive experience. At the time it may not seem so as cherished misconceptions and self-deceptions are unceremoniously smashed either by force of outer circumstance or through sudden inner revelation. As such a transit approaches, therefore, it is well to reconsider one's assumptions in advance in order to help smooth out the process. In the case of very inflexible individuals, this can be a time of devastating disillusionment or mental breakdown, whereas the more adaptable personality will respond in a more positive fashion.

On a more physical level, this transit can be a health hazard, bringing on maladies of a sudden nature such as accidents and heart attacks. Except where the cause of such a mishap is external and unavoidable, nothing is likely to occur if the basic health is sound.

Other hard transiting aspects of Uranus (square, opposition) will have a similar but much less noticeable effect. Reinforcing aspects (trine, sextile) will tend to be periods of positive change and discovery on a personal level that will occur in a very natural and easy fashion.

The Moon

The transiting conjunction of Uranus to the natal Moon will usually have an unsettling effect upon the personality, but often one which cannot be too well specified. It will tend to make the emotions somewhat jumpy and aberrant without any apparent reason. In individuals in which this is already a normal state, it can therefore cause considerable upheaval and can result in making emotional impressions or communications that may not be intended and can later cause regrets. This is particularly so because the effect is a transitory one and may frequently only be caused by irritating or upsetting external phenomena, after which the emotional state will return to its normal condition.

If the chart is heavily centered around the Moon, greater effects may be observed, such as have been noticed in the transit to the Sun. Yet these profound and long-lasting effects are less likely here than with the Sun and there is often no noticeable effect on the personality in the long range at all. It is, however, advisable to try to rein one's reactions in emotional situations, as the tendency to rashness that may be regretted is greatly heightened at this time. Basically, the Moon and Uranus have very little in common and therefore such transits may be generally expected to be of only momentary significance in all but a few special cases.

The same may be applied to other hard aspects in transits; the trine and sextile may go by with no noticeable effect whatsoever.

Mercury

Uranus has so much in common with Mercury that almost any transit will be beneficial, except in the most hidebound of individuals. The conjunction brings on flashes of inspiration and allows important intellectual breakthroughs and revelations that were not thought possible before. Only in very mentally conservative persons or persons of disturbed intellect might this cause upset through a forced change in view or bring on neural damage.

It can, however, bring a radical change in lifestyle as a result of intellectual decisions concerning how the life should be run. Where the natal Mercury is weak or afflicted, these decisions may turn out to be incorrect or their implementation may be effected in too extreme a fashion. In any case where a sudden decision or flash of inspiration may be likely, at least temporary moderation is to be recommended as Uranus does tend toward the extreme and in most life situations extremes do not have an altogether positive effect.

In general this transit may be much looked forward to and the understanding gained during it should be reassuring and long-remembered. Very often it will mark a turning-point in personal commitment and direction due to a profound change in outlook, usually a better understanding of the individual's mental capacities, inclination, style, and potential. Other hard aspects will have a less spectacular, but similar effect, while easy aspects will have a more gradual and evolutionary effect upon the mind.

Venus

Uranus is frequently connected to strong and/or unusual sexual desire and motivation. The Uranus transit, particularly the conjunction of it will have the tendency of changing and transforming the sexual nature. This does not mean that the individual will suddenly become obsessed with the subject or find his or her natural desires warped or distorted, but it will be a time of personal discovery in these areas in all but the most repressed of individuals. New attitudes will be formed that extend the physical and mental abilities to explore and enjoy sexuality which, if handled with tact, will only be for the good, though if not, can run to extremes.

Other elements of taste may also be affected at this time, such as the nature of the appetite and the kinds of foods one prefers as well as other personal tastes of an artistic or aesthetic nature.

An undesirable side of this transit can be the sudden change or interruption of the source of income, somewhat similar to the Uranus transit of the second house. In general, one may expect the alteration of those things in the environment which are nurturing or supportive in nature, though not necessarily in a negative fashion but rather in a stylistic one that may indeed be for the better.

Other hard aspects will bring similar, though less noticeable, effects and easy aspects will probably go unnoticed, though they may bring a certain outside recognition for originality in taste.

Mars

Mars represents the most physically aggressive side of Uranus and when Uranus transits Mars it will tend to emphasize the aggressive nature. During the transiting conjunction care should be taken not to overreact or become too emphatic in making a statement or point in social contexts.

For those involved in activities which require focusing of great energy, as in sports, this can be a time of great achievement. Even for more sedentary types, it may prove a time of significant reevaluation of the way physical energies are deployed so that they may not be wasted but used in a more careful and intelligent fashion. It would be wise for those not in the peak of health and training to avoid excess exertion as there may be a tendency to do so at this time due to increased energies and less control. Exercise and experiments with firearms or other dangerous and swift machinery should be avoided or done with extreme care, as there is a tendency for things to get out of control. On the other hand, because of this influence self-control can be well-developed because it will be most under stress during this transit. It is a perfect time for reason to master the mind and the body.

Other hard aspects, particularly the square, will have a more grating and irritating influence on energy expression, whereas easy aspects will tend toward greater and perhaps deceptive self-control.

Jupiter

The Uranus transit to Jupiter is often an excellent one, increasing creativity manyfold and providing new directions both internally and externally. The two planets are very compatible and although they tend to make sparks fly, the results tend to be beneficial due to the nature of Jupiter. If there is any warning to be made about what may seem to be nothing but positive new directions, it is that the transit is temporary and that the energy thus provided will not last forever. Rather, it must be gathered and directed in such a way that it may be most taken advantage of in the long run.

Where Jupiter is in hard aspect in the natal chart, then caution in launching into new situations is advised because of the extremity of the influence of Uranus. It will be easy to go too far in any new situation and thus lose the balance and advantage that was initially offered by it.

At best, this transit is the indicator of sudden and unexpected fortune, at worst it is the source of overstated optimism and the error in judgement that may accompany it. Only in other hard aspects, particularly the square, will it tend to promise more than it delivers and make the individual dissatisfied with his gains, even though they may be significant.

Easy aspects of Uranus will simply tend to encourage a flow of regular creativity which may be sufficient if a continuity of direction is already established but which will be of little help if that is not the case.

Saturn

The transiting conjunction of Uranus to natal Saturn is usually a period of considerable insecurity. The normal sources of support will often suddenly give way, but not infrequently to situations which provide even greater possibilities than were to be had before. For the conservative individual it will be a time of great stress, for the adaptable person it can be a time of great opportunity.

In either case, it will cause the individual to reevaluate the outer and inner sources of support and security and come to a clearer definition and understanding of them. Internally, this may result in considerably greater self-reliance. Reexamination of externals may make the individual more acutely aware of his dependence on factors surrounding him that have turned out to be more fragile than imagined.

This will always be a time for the immediate employment of inner resources in response to the vagaries of external forces. Ultimately, it is usually beneficial, creating both greater ability to rise up and meet circumstances and greater understanding of just how fickle even the most reliable of circumstances can be. The result is a more effective deployment of defensive perimeters and the elimination of those which liable one to harm from the unexpected.

The other hard aspects of Uranus will have only a minor unsettling effect, but the easy ones will have an excellent fine tuning effect upon the more precise areas of judgement in external affairs or any place where precision and exactitude of character is required.

Uranus

The transiting conjunction to the natal Uranus, or Uranus return as it is called, happens at age 84 and thus few live to enjoy it. Those who do frequently are not in good enough health to fully reap its benefits, either. Those that are in a position to benefit report that this is an exceedingly energetic and invigorating time in life and a kind of intellectual and spiritual rebirth. This may also perhaps be characterized by the fact that the same transit marks the end of a bilaterally symmetrical series of planetary cycles of those planets interior to Uranus in the solar system (for further thoughts on this, see my earlier work *Astrological Cycles and the Life Crisis Periods*). It is unfortunate that so few individuals are still in a condition to fully benefit from this transit at such a mature age.

The opposition, at age 42, is also associated with other bottoming cycles, and marks a general low point in most persons' lives when the readjustments of personality concept must be made as the result of the coming of middle age. For those whose main focus is interior, this is a very positive period; for more externally-oriented persons it can be a very upsetting time marking the loss of youth and the forced establishment of a new way of relating to the world. The square, at ages 21 and 63, mark the inceptions of legal maturity and incipient retirement and the concomitant responsibilities and adjustments they bring. Easy aspects seem to indicate periods of more natural and less demanding integration with society in general.

Neptune

The natures of Uranus and Neptune are decidedly opposite and may be characterized when in contact with each other as representing concrete reality as opposed to idealized or desired reality. Thus, the transiting conjunction of Uranus to Neptune is often an uncomfortable period in most persons' lives when harsh reality crushes out hopes and desires previously held and defended. It is a period when the individual is forced to get down to practical matters that frequently conflict with what he feels to be right and/or desirable but which must be dealt with nonetheless. It is most commonly distinguished by disillusionment and a resulting cynicism that takes some time for the individual to recover from.

While it is rare, this may be a time of sudden mystical discovery or revelation, particularly in cases where such experiences are brought on by the use of external means of altering the consciousness such as drugs or experiences which cause altered chemical states within the body, such as disease or trauma.

Ideally, such a transit might result in high mystical revelation as the most advanced nature of the two planets suggest, but that is seldom the case and more often has the effect of turning mystic to cynic. The reverse has been known to happen; cynics are prime materials for mystics.

Other hard transits will have a similar effect,

though less decided, except in the case of the square where the avoidance of drugs is to be recommended. Easier aspects make the integration of idealism and reality either temporarily easy or unnecessary.

Pluto

The transiting conjunction of Uranus to natal Pluto gives a better view of the necessity and techniques of controlling personal power as well as insight into the nature of the social realm. It will either be a period of the individual's subjection to the power and influence of others or the opposite. Each situation will usually give a permanent lesson in social relationships, since the transit will not occur again.

In any aspect, transiting Uranus will affect the internal ability to deal with others in a personal and psychological sense. This will always be revealing, but it may also be painful or disappointing at the same time as Uranus does not tend to be merciful when dealing with the psyche. Rather, situations of extremity and permanence may be expected which, if they are reinforcing as in the easy aspects, may give an undue sense of personal power and self-righteousness that could be later totally contradicted by reality. On the other hand, in the case of hard aspects, powerful confrontations on an inner and outer level may be expected. These will require the utmost in spiritual understanding so that they will not crush the soul, or be the breeding grounds for deep resentment and vengefulness. These negative feelings result in little good unless understood and transformed by understanding and forgiveness.

In any event, this transit is connected with generational trends, as the transit during one year may cover those born as much as a dozen years apart.

North and South Nodes

Direct transits (conjunctions) of Uranus to either of these points bodes little good and is definitely to be prepared for. In the case of the more benefic North

Node, it may mean sudden undertakings that require radical rearranging of the life to meet them. In such cases, only the most adaptable of individuals will obtain maximum benefit. For most, the advantage of opportunity and disruption that it provides will cancel out, leaving only a certain amount of confusion and readjustment.

In the case of the South Node, there is little to be done and results are usually unpreventable disaster of a rather sudden and upsetting nature. Of course, as with all transits, nothing at all may happen, but when it does it will seldom be enjoyable and will not be something that can easily be taken advantage of. If there is an explanation for this, it may lie in the realm of karma where such individuals just have it coming on a cosmic level and this Uranus transit simply means their time is up.

In either case, however, a distinct lesson is to be learned and the incidents involved do not go by without real learning from them. Here, truly, experience is the best teacher, though what kind of learning one may be forced into is unpredictable.

Aspects other than the conjunction will have little or no effect at all unless there is a mutually aspecting body on the node at the time or unless the node is natally conjunct another body. In both cases, the results will be due to and characterized by that other body.

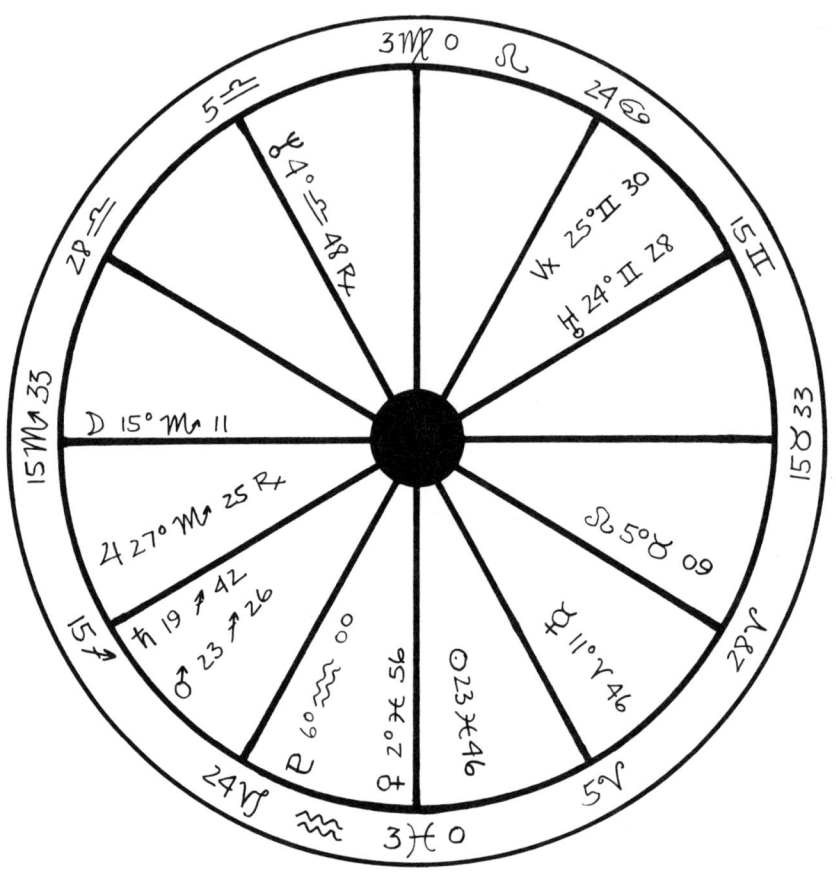

DISCOVERY OF URANUS

MARCH 13, 1781

11:00 PM

LONDON, ENGLAND

XVI
The Discovery Chart

After nearly 200 years of contemplating the position of Uranus in the horoscope, astrologers still find there is much to be discovered about its nature. Most of the information available about the planets has been derived from observation of its action within individual nativities and its effects when transiting, certainly reliable and time-tested methods.

Yet there may be another way to throw light on the meaning of the planet which requires less data and may also reveal some of its broader and more esoteric nature. That method is the examination of the positions of the planets at the time of the discovery of Uranus, when it first came into human consciousness, with particular attention to the degree symbolism of such positions and also the so-called "fixed" stars were present in those degrees.

Any such discovery chart must be somewhat speculative, as the exact time of sighting was not recorded, but we do know that it must have been sometime between 8:30 PM when twilight had entirely dissipated and shortly after 11 PM, after which a fairly bright waning gibbous moon would rise and discourage focusing on very faint objects. Since Herschel records that it was sighted on his last attempt of the evening, we may speculate that it was fairly close to 11 PM. Although the exact angles of the chart may be a bit uncertain, the planetary and moon positions may thus be extrapolated with fair certainty.

The relative positions of planets may give some indication of the nature of Uranus, or at least its future impact upon humanity. Uranus itself is in the Aquarian decanate of Gemini and is opposing Saturn, which it will come to replace as the ruler of Aquarius. It is directly aspected to the Sun (square), Mercury (quintile), and Mars (opposition), with all of which its qualities will become associated. It is not aspected to Venus, Neptune, the Moon, or Pluto, which might be considered logical, though it may be odd that Jupiter (only a wide applying quincunx) also is left out.

There is no earth in the chart, a hint at the extremity of the planet, and it is coming out of the hidden eighth house to replace Saturn in the very established 2nd, at least as far as rulerships are concerned. The probable certainty of Scorpio on the ascendant with the Moon is further indication of the sudden surfacing of something heretofore hidden in the depths of consciousness.

The T-square formed by Uranus opposing Saturn and Mars, both squaring the Sun, is most important. The coming of Uranus into consciousness was to usher in an age of revolutions, both political and technological, that was to have the most unsettling effect on established holdings (2nd house) and men's belief in the physical and political foundations of things (Sun, Pisces, 4th), all coming quite unexpectedly out of the blue (8th).

An examination of the degree symbols and coinciding fixed stars (adjusted for precession to 1781) also brings rich and meaningful rewards:

Uranus, in the 25th degree of Gemini, conjunct Betelgeuse (also Polaris, which is much further north, being the Pole Star, and of less significance probably). Betelgeuse, traditionally associated with Mars and Mercury, is a giant red star. It is considered to bring high military honors and great fame, particularly if associated with Jupiter (q.v. Chapter XIII) by favorable aspect.

Sepharial's translation of *La Volasfera* associates the degree with mental powers and ancient learning, while Jones says "here is the bending of all natural resources to the will of the individual." The image is one of cutting and trimming the environment to suit design. Other sources (Leinbach, Carelli) also associate the degree with intellect and crystallization of evolutionary form (Rudhyar).

The message of the degree may be said to be the crystallization of truth by the power of the will, qualities certainly associated with Uranus having both good and ill effects, depending upon their purity and motive. The degree is prominent in the charts of inventor-scientists Wernher Von Braun (Ascendant) and Igor Sikorsky (Mercury), who were both known for their stubbornness and tenacity as well as their inventive brilliance, as was Freud (Saturn).

The Sun, in the 24th degree of Pisces, conjunct Scheat. Scheat is considered by most an unfortunate star, bringing misfortune through accidents, explosions, wrecks, flooding, and other catastrophes of a Uranian nature. The MC of the first atomic test explosion was conjunct Scheat. Various sources attribute the qualities of Mars, Mercury, and sometimes Saturn to this malefic, and few have a good word to say about it. Its only positive influences seem to be when channelled through the mind, as in the case of Goethe and Nietzsche (Jupiter), or Rudolf Steiner and Victor Hugo (Mercury).

Cherubel describes this degree aptly as that of a person possessing great magnetic powers and the nature of a reformer who destroys evil, though Sepharial calls it a degree of very Neptunian sensuality akin to drug addiction. Jones, however, attributes to it inventiveness and independence, citing complacency and self-indulgence as negative qualities, which may unite the disparity of the previous sources. Jones and Rudhyar give the symbol as an island full of people in close interaction surrounded by a wide sea, a very Aquarian image where individuality must be carefully delineated in order to become independent and integrated at the same time.

Interestingly enough, both Leinbach and Carelli paint the degree as one of extreme sexual obsession, a trait for which Uranus is well-known. Neither has much good to say about this quality, though Carelli does quote its manifestation on a "higher plane" in the chart of Paul Ehrlich, who pioneered the war against syphilis. For good or ill, extremes of sexuality are certainly associated with the planet, though by no means are they attached to Uranus alone, sharing the distinction with Saturn and Pluto in particular.

The Moon, in the 16th degree of Scorpio, conjunct North Scales. The fixed star here at Uranus' discovery is associated variously with Jupiter and Mercury, but also with Mars. It is thought to bring good fortune, ambition, and permanent success, though this may be taken with a grain of salt, considering it was conjunct Mussolini's Ascendant. It is also, however, featured in the charts of Mahatma Gandhi (Mars) and Winston Churchill (Mercury). It is supposed to arouse spiritual and mental forces when favored and, in general, is said to have a positive influence on the mind.

Both Cherubel and Sephariel agree that the degree is one of open friendship, humanitarianism, and fiery devotion denoting spiritual loftiness manifested in concrete terms. Jones and Rudhyar see it as a degree of openness and faith, wherein the soul meets the world with the unsullied enthusiasm and acquiescense of a child. The image is suggestive of Christ's statement that one must become as a little child to enter into the kingdom of heaven. Carelli probably crystallizes the image at its highest level as a knight of the Holy Grail—energetically faithful and fearlessly loyal, but utterly pure of heart and without guile or malice in any form. It retains both the power and spiritual development of maturity without losing the purity and innocence of the childhood from which it sprung.

This characterizes the stubborness and willfulness of Uranus in its best light, where nothing but the truth is of any concern and no compromises are even seen or thought of, much less allowed. In the world of men, as Carelli points out, this tends to isolate the individual and take him on distant journeys in search of the light when staying home in the warm, if sometimes unchallenging, company of friends would be easier. The ultimate reward may be the greatest, but the path is seldom an easy one.

Mercury, in the 12th degree of Aries, conjunct Sirrah, also called Alpheratz. This fixed double star is supposed to bring fortune and popularity, having the nature of Jupiter and Venus and has the effect of bringing Uranus tendencies into public favor, as in the case of Thomas Edison whose Uranus was there and midpoint Mercury and Jupiter. It is also supposed to lend freedom and independence and would seem to describe appropriately the intellectual concepts of Uranus rather than the sometimes dire or stormy realizations it brings upon the physical plane. Its effect is to make these aspects more attractive as well, as in the case of Lenin (Mars) and Bismarck (Sun) or, less permanently, Nazi foreign affairs minister Von Ribbentrop.

Cherubel and Sepharial contradict each other, as is often the case, concerning the meaning of the degree, the former attributing willful eccentricism to it while the latter considers it sociable and conservative. In something of a combination of the two again, Jones and Rudhyar see it as a will to freedom in response to a higher cosmic order. This makes the degree radical on a lower and conservative clinging to inner principles resulting in a lonely but lofty existence, as in Carelli's image of an eagle in its mountaintop nest. Jones' and Rudhyar's symbol, however, is a flock of flying geese, a more Aquarian image. This conflict between the lone seeker after truth and the harmonized individual who believes he has found it while yet functioning with social structure might therefore be considered a particular characteristic of the Uranian mind.

Venus, in the 3rd degree of Pisces, conjunct Deneb Adige. Deneb yields a Mercury/Venus influence and is conducive to general success where the profitable use of ideas is concerned. Ingenuity and quickness of learning are implied, as with Hermann Hesse (North Node), and Da Vinci (Moon). But more rugged types may also be found in this connection accompanied by fame due to individual leadership or discovery, as with George Washington (Sun) or archaeologist Heinrich Schliemann.

Opinion is somewhat split on the degree symbolism here. Cherubel, Sepharial, Leinbach, and Carelli all agree that it is a degree of rugged individualism and reckless luck, sometimes inordinately tainted by uncontrolled physical appetite. Certainly where Venus touches Uranus in most charts these qualities may often be found, Uranus making Venus appetites offbeat and often extreme, though seldom ill-motivated.

Jones and Rudhyar see it as a petrified forest retaining the essential time-enduring qualities of culture and archtypical civilization. This image reveals a crystalized version of the essential needs, desires, and creations of mankind.

These may not conflict if taken at different levels: one being of the everyday nature, and the other in an historical or philosophical light. The former, more down-to-earth description will be for the majority the most appropriate, as Uranian desires do run to the coarse and eccentric, though good-natured—at least where one is dealing with the individual.

Mars, in the 24th degree of Sagittarius, conjunct Ettanin, or Acumen. This star is variously attributed the qualities of Mars, the Moon, Saturn, and Jupiter. It is associated with downfall in general, blindness, and social estrangement, though it can also enhance mental concentration and give an inclination toward esoteric studies. More successful examples of the better side of it are Sepharial (Asc.) and Da Vinci (Asc.), as well as Isaac Newton (Mercury).

As is the most likely outcome of the conjunction of Mars and Uranus, at least at first manifestation, this degree is associated with misfortune, accident, violence, early failure, suicide, or other self-caused demise. It is at best plagued with uncertainty and wandering and has an inherent unlucky instability. At least most indications and opinions seem to suggest this.

Jones and Rudhyar, as usual, have a more cheery and long-range view of it, representing it as a bluebird sitting on a cottage gate. It is supposed to portend the happiness and good fortune of the well-integrated soul, far from the shiftless wreck of other authors! One may only surmise that this is an end devoutly to be struggled for through resisting the harsh and rash effects of Mars-Uranus at a lower level of consciousness, i.e. the one most of us experience everyday.

Perhaps the moral may be taken from the explosive nature of the combination to begin with. The basic quality is one of aggression and destruction, but refined and sharpened to a laser-edge and used in the hands of a master, those same energies may be put into beneficial service for all.

Jupiter, in the 28th degree of Scorpio, conjunct Yed Prior and next to it Akrab or Isidis and Graffias. A star of the nature of Saturn and Venus in the former and Saturn and Mars in the latter, all are connected with revolution and "shameless immorality" both of a sexual nature and also concerning falsehood and treason. In higher development, mental capacities are indicated, but in lower manifestation there is indication of physical injury or disability. Elizabeth Taylor has her Ascendant here and, as is well known, has been accused of most of the negative qualities above mentioned.

This seems to be a degree of extreme emotional commitment. In its higher light it can be reflected in great personal loyalty or religious faith. In its lower aspect it might be expressed as revenge, jealousy, cruelty, or treachery. In either case, ability and effectiveness, together with planning are indicated, thus causing the effects to be the best or worst because of it. Even the more positive views of Jones and Rudhyar follow the same theme, stating it as the lesson of overcoming personal pride and the holding to the lasting principles of unity and togetherness, despite the inconsistencies of society. In higher development, this might engender unswerving faith and allegiance to a positive cause, while in lower manifestation it might cause the personality to plot and make war with great tenacity upon anything that contradicted his immediate beliefs or goals. Certainly the pairing of Jupiter with Uranus would bring great outpourings of energy with basically good intent, even though the effect might not always be benefic either in hindsight or from others' points of view.

Saturn, in the 20th degree of Sagittarius, conjunct Lesath. Lesath is of the nature of Mars and

Mercury and in medieval times had the reputation of inclining the native to be torn apart by wild beasts. Even now it is associated with accidents, operations, and other physical mutilations, deformations. Notably, it is the sting of the constellation Scorpio and is as such associated with poisons and physical betrayal.

As a degree by itself, it fares much better and is generally considered social in nature, concerning the duties and responsibilities of man to the social structure around him. Jones and Rudhyar use the symbol of villagers cutting up ice from the river to store for use in the summer, but in a convivial and enjoyable fashion where all are glad to serve the good of the community. There is considerable artistic and sometimes medical talent here, and there may also be a certain restlessness and desire for social change for the good of all as in the case of Marx (Uranus, ruler of Asc.).

Throughout the desire is both for stability and a certain precision of competence with which to maintain it. Certainly the natures of Saturn and Uranus blend rather well in a social context, particularly as subsequent rulers of Aquarius, and the two are often found together in matters concerning the technological support of science and industry. Perhaps some of the physical destructiveness of Lesath and this fairly mild and cooperative degree may meet in the basic urge to manipulate and physically reshape the body or the environment, whether through medical, social, or accidental purpose.

Neptune, in the fifth degree of Libra, conjunct Vindemiatrix and Caphir. Vindemiatrix is said to represent Ampelos, son of a satyr and a nymph and favorite of Bacchus, called the Gatherer of Grapes. Variously attributed the qualities of Venus, Mercury, and Saturn, it is associated with falsity, deceit, and other of the poorer qualities of Neptune. Caphir, on the other hand, is given to Mercury, Venus, and Mars and called An Atonement Offering and The Submissive One.

It seems to share the positive qualities of Neptune such as a refined, courteous, and lovable character combined with prophetic instincts.

As a degree, 5 of Libra is generally looked on as one of revealed inner knowledge and the inspired transmission of spiritual wisdom. It is not, however, one of simply passive receptivity to the Word, but rather one which in the fire of inspired belief is ready to take up the sword and conquer all for the Truth. The result can sometimes be danger and personal harm from fighting too strongly for ideals at inappropriate times, or holding too dogmatically to principles which are ill-defined. At its highest level it combines both the faith and refinement of Neptune with the strength and specificity of Uranus in the form of the truly revealed Word. At more mundane levels, however, the less pure influences of both planets are felt. This gives both great motivation and great frustration at the same time as one works against the other, leading to an impetuous and sometimes dangerous will on one side and helpless entrapment on the other as, to quote Charubel, "a man in a bog."

Pluto, in the seventh degree of Aquarius, with no fixed star conjunction. It is a degree with two common interpretations, one lower and one higher. On the low side it represents hidden power and strength which is not self-aware and therefore, like many of the powers of the subconscious mind, is either used for destructive purposes or is ignored entirely. Because of the lack of self-understanding, it creates great vulnerability despite its attributes of power and strength. It leads toward impulsiveness marked by convictions that cannot be realized in a peaceful or constructive way because of lack of insight concerning their applicability. Cherubel, Sepharial, Leinbach, and Carelli all view the degree as one of great and sometimes overwhelming or violent potential thwarted by lack of proper understanding and thus unnecessarily turned to destructive purposes.

Jones and Rudhyar, as may be expected, view the higher and more positive possibilities of the degree, seeing it as one of cosmic transmutation and evolution, as represented by the symbol of a child being born out of an egg. It represents a totally fresh start and new direction, not developed as a result of earlier changes but grown up totally new and different and unrelated to what had gone before. It is said to possess tremendous resourcefulness and may be viewed as a completely self-aware version of the less constructive image painted by the other authors cited. In this case, the power to change and transform is used with clear vision thus laying the foundations of the new without the destructive effect this energy can have when not fully formed or directed.

North Node, in the sixth degree of Taurus, conjunct Schedir and El-Nath. Schedir is given a combination of Venus and Saturn, in which seriousness and personal enjoyment are combined. In good aspect, this can produce pleasure of real meaning untainted by shallow frivolity, but in ill aspect it brings extremity of appetite and desire that can cause harm from overindulgence. El-Nath is associated with Mars and Saturn working against each other and is generally considered quite demonic in its effect, bringing material danger, cruelty, and madness or death. Rudolf Hess had Sun here and Mata Hari had Neptune, for example, though it can work less disastrously as with Rudolf Steiner (Mars).

In general the degree is viewed as one of great intellectual perception and a diverse and incisive mind. It also indicates an Aquarian nature with which the individual can make many different sides of a situation come into cooperation so that new advances may be made through collective effort and understanding.

On quite another level, it is also viewed as a degree of excessive passion and sexual extremes. This may either be manifested in complete licentiousness or aberrant behavior, or, oppositely, the complete withdrawal from the physical in order to give full attention to the application of the mind. Both conditions are typical of Uranus' first approach to new undertakings (North Node) and, usually, time is required before the higher manifestation of cooperative multiplicity can be gained.

The South Node, in the sixth degree of Scorpio, with no fixed star conjunction. In a mundane sense, this is a degree of stubbornness, determination and the careful building of an edifice through great care, attention to details and exactitude of execution. This can be employed for either good or ill. It is necessary here to see that the details can become an obsession, obscuring all else. When ill-aspected, there is a hint of spiteful duplicity and treachery, which may be considered an offshoot of the obsessive personality.

Jones and Rudhyar, however, depict the degree using the symbol of a gold rush, with men rushing to seize the benefits of a radically changed world and social situation. It is a degree of ambition accented by the need for rather spectacular and radical change to new and different values and social situations.

Considering the South Node to represent the past situation and what is inherited from it, this may fairly well represent the situation at the time of the discovery of Uranus. The period marked the end of the gradual growth of technology and governmental form and signaled the beginning of radical social and technological reorganization which brought on a completely new order. The mass of society seized upon the new technology and social structure in just the manner of a gold rush. This left the world permanently and radically altered.

Before one may consider the symbolism of the angles of the Uranus discovery chart, it must be remembered that there is not a settled "exact" time from which to calculate them. In such a case, some sort of rectification must be employed to deduce the time based on traditional astrological principles and methods. Without using the degree symbols themselves as a basis for rectification, which in this case would obviously be improper, and having no external data to go on, there is only one avenue of logic left, given the approximate time of 11 PM or shortly before. Tradition and experience indicate that important events occur when celestial bodies are closely conjunct the angles. If we use a time of about seven or eight minutes before eleven, we find this pattern on all three angles (the third being the prime vertical), something that no other time yields during the period Uranus was visible that night. This places the Moon conjunct the Ascendant, not yet visible about the horizon because of the parallax, Uranus is conjunct the Vertex, and the fixed star Alioth in the Great Bear (we shall see the significance of that later) is at the Midheaven. From the point of view of standard rectification, nothing could be more ideal, though whether it is in fact the time of discovery is naturally still speculative.

Using the positions based on this time, we may then observe the degree symbolism and see if it falls as neatly in line with the characteristics of Uranus as all the previous positions have. If they do, we may be content to look no further:

The Ascendant, in the 16th degree of Scorpio, is identical to the rising Moon, so no more need be said about it except to give double emphasis to a degree that has considerable symbolic impact already, as represented by the purest seeker of Truth, the Knight of the Holy Grail.

The Midheaven, in the 4th degree of Virgo, conjunct Alioth. Alioth is far from the ecliptic and is not given great significance in traditional astrology except some mention it as having a Martian quality. In esoteric thought, however, it has great significance as one of the main spiritual ray sources of the solar system. The ray it emits, the third or intellectual ray, is in fact esoterically ruled by Uranus! One could not ask for more than that . . .

Like Uranus, the degree symbolism is one of total extremes or the blending of them in total harmony on a higher plateau. Carelli's image is that of a steam roller and he quotes Isaiah XL, 3-4: "Prepare ye the way of the Lord, make straight in the desert a highway for our God;

"Every valley shall be exalted and every mountain and hill shall be made low, and the crooked shall be made straight, and the rough places plain."

But in the same breath, he also quotes Attila the Hun: "No grass grows where my horse has trodden." The degree is one of totality, whether for good or ill. It has in it the inexorable seeds of destiny and change, as also represented by the Tarot trump "The Chariot."

To Jones and Rudhyar, the degree is one of Aquarianism in the extreme, symbolized by the brotherhood of black and white children playing together. It is the human extremes adjusted to live in harmony and equality in the so-called Aquarian Age. Whether this ideal may be realized or whether the more unruly and lower aspects of Uranus will continue to hold sway is a matter of debate that is not likely to be resolved by man alone in the near future. Suffice it to say, however, that the symbolism fits and is appropriate.

Finally, the *Vertex*, in the 26th degree of Gemini, conjunct Uranus itself and the fixed stars Polaris and Betelgeuse. These stars and their symbolism were previously described (under *Uranus*) and therefore bear only a reemphasis, particularly in regard to the long-range effect Uranus has in military operations, since the Vertex is concerned with fateful turning points and these are of the most prominent type among the affairs of men.

The degree symbolizes, in its lower form, willfulness and adherance to dogma, particularly dogma of an individual's own making. Thus, it tends to be a degree of strife and conflict, pitting one person's or culture's view of truth against another, the weak link in the scripture of almost every revealed religion. In its higher form, it represents (as in Jones and Rudhyar) the crystalized quintessential truth of things, the essence from which all variations depart. Here we find Uranus at its purest and most laconic, being the kernal of truth whose systematic embellishments each culture calls Religion or Science, depending upon its style and inclinations.

One might carry the inspection a step further into more experimental ground and look at the *East Point*, about which not all that much has been determined. The general belief is that it has to do with the direct physical state and health of the body. In the Uranus discovery chart it is in the eighth degree of Sagittarius, conjunct Rastaban. That star is noted for various accidents, illnesses and sudden disasters which are certainly notable in the transits of Uranus, particularly to the Ascendant. The degree symbolism is equally Uranian in context, typified on the lower level with the all-or-nothing risk of the extreme gambler or the person in a desperate situation and, on the higher level, of the alchemical transmutation by fire—the initiation of the spirit through total conflagration. Indeed, this is suggestive of the fiery raising of the *kundalini* in the initiate which can either enlighten or destroy, depending upon the preparedness of the individual.

What is most noticeable from viewing the fixed stars and degree symbols involved in the Uranus discovery chart is how well and thoroughly it ties in with the observed effects of the planet in the two centuries since it was discovered. It would appear that the basic qualities of Uranus, as far as astrology is concerned, could have been gleaned by the aware observer immediately upon its discovery, simply by the analysis of that chart. If that is the case, then the principle that the nativity of a thing or person (being that moment when it comes into human awareness) does indeed describe its nature in symbolic terms, as traditional astrology alleges. In addition, the observed effects of Uranus also tend to confirm the validity of degree and fixed star symbolism, particularly since the degrees and fixed stars are not interlocked due to precession and yet coincide so neatly in the case of the Uranus chart.

More ominous still or reassuring (depending upon your point of view), is the resulting conclusion that may be made: despite all human efforts, Uranus could not have come into human consciousness at any other time (within a four minute tolerance) than it did. Only the exact time it appeared could possibly describe it so well, backed by astronomical odds against any other planetary and degree combination coming up with the same thing. If it had come at any other time, its description and birth and thus (perhaps) its described nature and effects would be significantly different. This uncomfortably deterministic line of argument may bear some refutation, but it is somewhat disturbing nonetheless, at least as far as man's desire for "free agent" status on a cosmic scale is concerned.

Whether one chooses to believe that Herschel was propelled onto his roof that evening by cosmic forces, or that he just stumbled up there basically by accident or, indeed, that he virtually determined the course of the future by his acts that night, it cannot be denied that the descriptions of the fixed stars and degree areas do indeed describe the basic nature of the planet, a technique that might be of real value in establishing and delineating the qualities of more recently discovered heavenly bodies such as Neptune and Pluto or other as yet unobserved satellites of the Sun. Indeed, it might be a major source of interpretation of the yet entirely undeveloped possibility of weighing the effects of X-ray sources, quasars, pulsars, and other strong-radiation objects which have not been considered previously because of their relative invisibility to optical observation.

Uranus itself has been the subject of much observation of a non-material sort which, because of its basically unsubstantiated nature, remains speculative. These observations are those of the so-called "esoteric" persuasion, of which large-scale proof is yet to be provided. Yet, it must be remembered, that the very nature of Uranus is that of revealed and incontrovertible truth that is no less true before than after its general popular substantiation. Therefore the esoteric meanings of that planet, which do in many ways seem logical and proper extrapolations of its symbolism in a further context, may indeed be nothing less than the incontrovertible truth, at least as seen in "future hindsight." Some of these esoteric interpretations, as given almost exclusively to members of the Theosophical Society and their alleged adherants, are given in the following chapters.

XVII
Uranus in Esoteric Astrology

In order to outline the view of Uranus as seen from esoteric astrology, a few words about the field itself are needed, for it is entirely different from everyday traditional astrology.

Esoteric astrology as it is understood and practiced today is a relatively new pursuit, dating back only to the late nineteenth century, although planets have been said to have special secret meanings long before esoteric astrology became codified at that later time. It is of limited use to most astrologers as it does not deal very much with the physical world but rather with the non-physical universe claiming to be not the astrology of everyday matters but rather the astrology of the soul.

Unlike traditional astrology, whose tenets have grown up through the collective observations of many thousands of practitioners over the centuries, the structure and principles of modern esoteric astrology are strictly revealed by a single source, a group of non-material entities called the Great White Brotherhood, of which one, a Master Djwal Kul, has given primary attention to setting down on paper the rules and regulations of esoteric astrology.

The beginnings of esoteric astrology were revealed to Madame H.P. Blavatsky in the late 1800's, but the bulk of the material was given to Alice Bailey who has set it down in her 700-page *Esoteric Astrology*. Virtually all other contributions to esoteric astrology have been based on or extrapolated from that massive work. The system is an integral part of the much larger belief system called Theosophy, which Blavatsky and Colonel Olcott founded in 1875 and whose tenets are probably best explained not by Blavatsky herself but by C.W. Leadbeater and Annie Besant, who had a more succinct command of the English language.

Any attempt to understand (what is called) esoteric astrology without first becoming quite familiar with Theosophy is entirely hopeless. Indeed, even with that familiarity it is a difficult and often confusing task. The multiplicity of overlapping and sometimes contradictory meanings attributed to the planets and signs may well be of profitable use to someone who has or is achieving substantial first-hand experience of the non-material planes of existence and those who are reputed to dwell upon them, but most will find the whole subject more confusing than helpful. It is probably more profitable for the individual seeking to develop in that direction to make a serious and detailed study of Theosophy first and try to develop some first-hand experience in the spiritual world before delving into the more arcane side of astrology. Otherwise even a successful memorization of the many sets of rules therein will have little real impact and will tend to be words without substance.

Describing the esoteric meanings of Uranus out of context without first explaining the basic tenets of esoteric astrology in general would be entirely useless, while properly and sufficiently explaining these principles would make a book in itself, as indeed it has already. A few details, therefore, will have to suffice, and the reader is referred to Bailey if an in-depth study is desired.

First, esoteric astrology does not look at the planets as mere physical bodies in space whose gravitation or magnetic alignment effects matters on earth as traditional astrology does. Rather, it considers each planet a living spiritual entity that sends out varying forces on the spiritual as well as physical plane. Some of these planets, indeed, have no physical existence at all and exist entirely on the higher planes such as the astral, mental, causal, etc. There are a number of sets of these all whirling around the Sun, making the solar system quite crowded if all were taken into account. Esoteric astrology, however, uses only thirteen, seven so-called "sacred" planets, which have reached a certain level of spiritual achievement and evolution, and five "non-sacred" planets which have not. Uranus is considered one of the sacred planets.

Each of the seven sacred planets serves as a channel for the one of the seven spiritual "rays" which emanate from the seven stars of the Great Bear and which are responsible for the primary motivating spiritual power in our solar system. Of these seven, Uranus rules either the third, or intellectual, ray in its esoteric meaning or the fifth, or scientific, ray in its more mundane meaning. If all this begins to sound a bit far-fetched, remember that the third star of the Great Bear was overhead at Uranus's discovery, and coincidences like that are hard to come by.

The primary effect of Uranus in esoteric astrology, as in ordinary astrology, is illuminating rays as far as the development of the soul is concerned. It is thus associated with the level of spiritual progress where the kundalini fire is roused from its sleep at the base of the spine and brought up through the spiritual focus points of the body (the chakras) and released out the top of the head, giving the native the powers of spiritual sight and clairvoyance, among other things. All descriptions of kundalini-raising depict it as being quite violent and Uranian in nature, and if the individual is not prepared to handle it, it can result in permanent insanity, neural damage, or heart failure, so powerful is the vortex of energy that swirls up the spine at the time. Theosophists maintain that it is not an experience to be taken lightly and should not be attempted without assistance and guidance of more evolved helpers. Certainly it is a risky business when not fully controlled, as various testimonies, particularly that of modern writer Gopi Krishna, have noted. Certain practices of yoga, hatha yoga in particular, are said to bring it about, but without proper spiritual development to match, it is a dangerous game by all accounts.

Because of its connection with the Sun, Uranus is also connected with the sign Leo, said to represent the first stage of general spiritual initiation, something that is about to begin happening on a wholesale level as the Aquarian Age unfolds. It will be through the influence of Uranus and the higher qualities of Leo that this general illumination will come about. On an individual level, Uranus is said to influence and rule the fifth, or final, spiritual initiation at which time the native through his own efforts breaks free of the wheel of karma and is ready to move on to other areas of universal endeavor not necessarily connected with the physical plane or the human race.

Uranus is also said to be the esoteric ruler of Libra, probably essentially because both have to do with total truth and order in the universe, Uranus through revelation and Libra through action. In addition, in some instances Uranus is also linked with the seventh, or ceremonial, ray, particularly now when that ray is supposed to be becoming much more active as we move out of the influence of the devotional sixth Piscean Age ray.

In fact, in esoteric astrology, the rulerships of all the planets and signs shift depending upon which level of spiritual development or scale in creation is being discussed, and much further discussion of Uranus in this respect is futile by itself. It must be taken in the context of the particular level and direction of development of which there are many listed in Bailey both for human and spiritual or non-human entities.

If any synthesis of the many esoteric meanings of Uranus may be made, it can be said it has the primary qualities of spiritual fire which burns impurities out of the soul and alchemically transmutes the inner being from dross to fine matter in an overwhelming and sudden manner. It has none of the slower qualities of growth and development attributed to the other planets. Rather it is an all-or-nothing planet, as it is in traditional astrology.

All of this lore, it must be remembered, has come in quite Uranian fashion from the mouths of beings of higher spiritual evolution which seldom manifest themselves in matter. They are said to be the ones responsible for the formation of the Theosophical Society and early writings from that organization speak of considerable concrete contact with these so-called Masters during which time the rather detailed principles of Theosophy were set down.

Except for occasional alleged individual contacts, these entities have fairly well withdrawn themselves since early in the 20th century. This, of course, severly limits the gathering of further information upon the subject until such a time as they choose to be seen again or men in general are more conversant on the higher planes than they are now.

According to their parting messages, however, both situations are imminent and as we come closer and closer to understanding the real, tangible nature of "occult" phenomena through systematic research and investigation, contact will again be made, and this time it will not be quite so one-sided. Indeed, it is said that several of the Masters will make full-time incarnations, borrowing the bodies of pupils or other suitable people, and will begin to make the spiritual world a matter of real everyday fact and experience instead of one based on mystery and speculation. All this is in line with the doctrine that after many ages of spirit becoming tied up with matter, it is finally beginning to pull itself out so that it may move more freely in and out of the physical plane.

All this seems quite doctrinaire, but the same writings stress that those who wish to learn more of the subject should not take anyone's word for it, even the words of the Masters themselves. Rather, students are urged to confirm everything by investigation and personal experience. If there is a moral for Uranus in any of this, it is that the systematic exploration of unknown "occult" affairs that is seeing so much growth currently is quite the proper direction to take. It is time to make the matters of the spiritual world, life after death, karma, reincarnation, soul travel, and clairvoyance, to name a few, areas of concrete Uranian fact and reality no longer subject to the distortions of faith or blind belief.

Encouraging signs in current research show that this is not an unreasonable goal and that it may be attained with time and patience with or without direct intervention and help from higher beings. The Truth, concrete and absolute, is there for the asking if enough careful and intelligent investigators are around to dig it up. In this respect, it may be said that Uranus, the normal ruler of Aquarius and the Aquarian Age, is the motivating planet for man when he strives to see things for himself, taking nothing for granted or on belief. For thousands of years man has seen occult affairs through a glass darkly, but he is currently about to see them face to face, largely through his own efforts. Christ said "You shall know the truth and the truth shall set you free," and therefore there's no reason to take His or anyone else's word for it until we indeed do know the truth. The very scientific investigative process whch initially seemed to deny religion and the spiritual life are, in time, learning how to find direct evidence of it and confirm it in a way that makes it tangible and accessible. Now contact with the higher levels of our being is no longer a matter of faith, accident, or unknown influences.

One facet of Uranus is direct spiritual revelation, but that is only one side of the matter. The other side of Uranus is careful, precise investigation which draws out incontrovertible truth. It is likely that both of these sides of the nature of the planet must become more fully developed before we find ourselves coming to a significantly clearer view of occult matters.

Thus, we have a revealed side of esoteric astrology set down, but without its confirmation in a more concrete sense it can have little meaning or use to the world at large. It remains, perhaps, a pointer toward astrological meaning and interpretation to come, but until man has a more physical grasp of the basis upon which it is founded it will remain for most a promising enigma at best.

With the developing of traditional techniques and their application to non-material phenomena, however, the two forms of astrology may eventually blend and we will no longer have to divide the world into upper and lower modes of truth as there will be a thorough understanding by all of the way things really are, whatever that may turn out to be.

Hopefully, that time may not be far away, and if the Theosophists are to be believed, along with the countless other sects who are predicting an imminent Coming, the time is just around the corner. But what is important, however, is not getting it done quickly but rather getting it done well and it will not serve the intelligent astrologer to wade too deeply into esoteric astrology until he has a firm basis of experience for it. And when he does, it is not likely that he will require a book from which to learn it.